AF823487

Interrogating Psychiatric Narratives of Madness

Andrea Daley · Merrick D. Pilling
Editors

Interrogating Psychiatric Narratives of Madness

Documented Lives

palgrave
macmillan

Editors
Andrea Daley
School of Social Work
Renison University College
University of Waterloo
Waterloo, ON, Canada

Merrick D. Pilling
Women's and Gender Studies
in the School of Social Work
University of Windsor
Windsor, ON, Canada

ISBN 978-3-030-83691-7 ISBN 978-3-030-83692-4 (eBook)
https://doi.org/10.1007/978-3-030-83692-4

© The Editor(s) (if applicable) and The Author(s), under exclusive license to Springer Nature Switzerland AG 2021
This work is subject to copyright. All rights are solely and exclusively licensed by the Publisher, whether the whole or part of the material is concerned, specifically the rights of translation, reprinting, reuse of illustrations, recitation, broadcasting, reproduction on microfilms or in any other physical way, and transmission or information storage and retrieval, electronic adaptation, computer software, or by similar or dissimilar methodology now known or hereafter developed.
The use of general descriptive names, registered names, trademarks, service marks, etc. in this publication does not imply, even in the absence of a specific statement, that such names are exempt from the relevant protective laws and regulations and therefore free for general use.
The publisher, the authors and the editors are safe to assume that the advice and information in this book are believed to be true and accurate at the date of publication. Neither the publisher nor the authors or the editors give a warranty, expressed or implied, with respect to the material contained herein or for any errors or omissions that may have been made. The publisher remains neutral with regard to jurisdictional claims in published maps and institutional affiliations.

Cover credit: oxygen/gettyimages

This Palgrave Macmillan imprint is published by the registered company Springer Nature Switzerland AG
The registered company address is: Gewerbestrasse 11, 6330 Cham, Switzerland

Foreword

Every so often a new book is released that has the potential to shake up a research field with its paradigm-shifting revelations. *Interrogating Psychiatric Narratives of Madness: Documented Lives* is such a book in its critical analysis of the power of language and ideology to define truth. Its contributors, a team of mad studies scholars and qualitative researchers, present a landmark study that examines and highlights the ways in which social marginalization, whether due to socioeconomic status, gender, race, or sexual orientation, operates within the medical chart by ultimately defining reality. They show us the power of expert narratives to fortify and reproduce dominant ideologies of social control. This book is an important contribution to the field of critical mental health. It deepens our intersectional understanding of institutional representations of distressed persons highlighting psychiatric oppression as a serious and underrecognized social injustice that contributes to health inequities in Canada.

Methodologically unique, empirically rich, theoretically sophisticated, and analytically robust, this critical research questions the official narratives, discourses, and ideological underpinnings at play in the textual representation of psychiatrized persons, unveiling the semiotic strategies involved in the mortification of persons undergoing total institutionalization. As a ground-breaking study of psychiatric documentary systems, *Documented Lives* explores how medical texts function to promote psychocentrism, the dominant societal view that all human problems are innate pathologies of the individual mind or body, rather than responses to a pathological society built on social, political, and economic inequalities and injustices. This collection of essays reveals the deadly knowledge deficits of psychocentrism that prevent mental health professionals from understanding the impact of social factors and forces, thus compromising healthcare treatment for socially vulnerable persons.

The authors demonstrate the power of expert narrativization in representing patients in particular ways via the "medical chart," a primary site of meaning making that function to make up patients according to the interpretation of experts. Institutional agents can maintain and reproduce social inequalities by exerting power, using force, and affixing disease labels through various processes, functions, dynamics, norms, values, and relations intrinsic to psychiatric institutionalization. The dominant ideological assumption is that doctors are scientific experts of the mind who heal, treat, or cure ailments, and who operate according to their pledged oath, "to do no harm." The critical scholarship presented in these pages offers us another perspective, questioning largely presumptive claims to objectivity, neutrality, and knowledgeability.

Critical analysis of medical charts, bureaucratic records, and psychiatric evaluations of people in distress reveals the dominant social relations at work behind so-called neutral and impersonal scientific practices. The language and status of science are invoked to mask subjective clinical evaluations that become objectified forms of ideology in charts, reducing sufferers to mere case files or diseased entities. Often perceived and labelled deviant, brutalized, stigmatized, dehumanized, objectified, pathologized, infantilized, and inferiorized in and through the psychiatric text, patients are narrativized in derogatory ways. The descriptions of these patients often smack of a harsh judgmentality, disrespect, and outright disdain, rather than empathic, competent medical attention and the duty of care expected of the psychiatric institution in the twenty-first-century society.

The examination here of the psychocentric attitudes, perceptions, and interpretations contained within the medical chart offers readers a study of "psychocentrism in action." Rather than understanding the contributory role of inequalities in poor mental health, psychocentric ideologies entail focusing on individual flaws alone, promoting a human deficit model while obscuring the social causes underlying much mental distress. A broader, sociological understanding is thus erased within the medical text that constructs an abnormal individual that can be fixed with a pharmaco-rationality above all else. The medical chart thus operates to largely invalidate the social context, social history, social relations, and social experiences contributing to the experience of distress, thereby rendering invisible the effects of harmful social ideologies.

Documented Lives challenges institutional record-keeping, raising important questions regarding the conflicting interests of the different stakeholders, the quality of mental health care services, the iatrogenic and nosocomial risks associated with hospitalization, the lack of adequate trauma-informed practices, knowledge deficits of the well-established social determinants of mental distress and other significant epistemic holes, the lack of diagnostic evidence in the form of objective tests and measurements, and the subjective nature of expert perception and interpretation. The study also emphasizes the ethical responsibility of psychiatry to address issues of social harm, violence, and inequalities, both within and outside the institution that contribute to poor mental health outcomes.

Isolation, fear, stress, anxiety, anger, depression, alienation, self-harm, addiction, and suicidality must be placed within the context of social life and systemic structural oppressions if we are to overcome the social injustice of psychocentrism, systemic ableism, and the sanism that blames and shames those struggling with poor mental health. Psychocentrism frames those not "resilient" or immune to the micro and macro strains of social life, especially those embroiled in chronic conflicts and struggles based on age, sexuality, socioeconomic position, gender, (dis)ability, familial support, educational level, employment status, religious beliefs, racialization and ethnicity, and political affiliation, as well as traumatizing social experiences such as sexual violence, harassment, mobbing, and other serious forms of crime victimization.

Medical chart writing might unintentionally contribute to patient trauma, indignity, and injury by upholding rather than challenging harmful stereotypes and inequalities plaguing neoliberal society. Psychocentric attitudes and beliefs, discriminatory social and institutional norms and policies governing psychiatric inpatient wards mutually reinforce each other in a broader constellation of social relations of power and inequality. *Documented Lives* signals the urgent need to shift away from coercive and psychocentric psychiatrization toward a progressive, socially informed, rights-based approach that requires trauma-aware and trauma-competent healthcare providers. Systemic reformation is required to safely attend to those seeking relief in humane and respectful ways, rather than exacerbating or contributing to the chronic or acute mental distress of vulnerable persons who deserve life-affirming expertise above all else.

February 2021

Heidi Rimke, Ph.D.
Department of Sociology
The University of Winnipeg
Winnipeg, Canada

Acknowledgments Dr. Andrea Daley and Dr. Merrick D. Pilling extend their gratitude to Opal Sparks and Dawnmarie Harriott for their support of the project, key insights, and contributions to analytical themes. Their rich and critical perspectives on psychiatry, the mental health system, and patient experiences strengthen the book in innumerable ways.

Contents

Notes on Contributors

Lucy Costa is a Deputy Executive Director of the Empowerment Council, an independent service user rights-based organization in Toronto, Canada. She works as an advocate promoting the rights of mental health service users/survivors, as well as encouraging critical analysis about service user inclusion in the mental health sector. She has developed education curriculum for many stakeholders, including psychiatry residents at the University of Toronto. She has written a number of articles and blogs and is co-editor of *Madness Violence and Power: A Critical Collection* as well as co-editor of a special edition of the *Journal of Ethics and Mental Health.*

Andrea Daley is a Professor at the School of Social Work, Renison University College, at the University of Waterloo, Canada. She has published on social justice issues, including those impacting sexual and gender minority communities with a focus on access to equitable and good quality health care; lesbian/queer women's experiences of psychiatric services; and gender, sexuality, race, and class and the interpretative nature of psychiatric chart documentation as it relates to psychiatric narratives of distress. She practices critical research methods to engage politics of knowledge building with communities toward the goal of social transformation. Andrea's work as a community mental health worker has been particularly influential in shaping her commitment to addressing the issues of inequity and oppression within the context of psychiatric care. She is a co-editor of *Madness, Violence, and Power: A Critical Collection* (University of Toronto Press).

Margaret F. Gibson is an Assistant Professor in Social Development Studies and the School of Social Work at Renison University College, at the University of Waterloo, Canada. Her research focuses on how texts, institutions, and encounters reproduce ideas of difference and exclusion, particularly with regard to sexuality, gender, kinship, and disability. Her publications include work on the history of psychiatry and related fields, including research about "female inversion" and "feminine boys," as well as on contemporary interactions between systems and LGBTQ+ parents. Her current work includes an ethnographic investigation of how people are using the idea and language of neurodiversity.

Merrick D. Pilling is an Assistant Professor of Women's and Gender Studies in the School of Social Work at the University of Windsor. His work employs Mad Studies and intersectional, feminist approaches to understanding mental distress and experiences of violence and trauma. His publications include chapters in the anthologies *Containing Madness: Gender and "Psy" in Institutional Contexts* (Palgrave Macmillan) and *Madness, Violence and Power: A Critical Collection* (University of Toronto Press). His book *Queer and Trans Madness: Struggles for Social Justice* is forthcoming with Palgrave Macmillan in 2021.

Lori E. Ross is an Associate Professor in the Division of Social and Behavioural Health Sciences, Dalla Lana School of Public Health, University of Toronto. She is an established interdisciplinary scholar in the field of sexual and gender minority (2SLGBTQ+) health, with a particular focus on mental health. Her current program of research focuses predominantly on qualitative and mixed methods investigations of health experiences for members of the 2SLGBTQ+ communities, with a focus on examining the contributions of intersecting structural oppressions to these experiences.

Juveria Zaheer is an Assistant Professor in the Department of Psychiatry, University of Toronto, and a clinician-scientist at the Institute for Mental Health Policy Research at the Centre for Addiction and Mental Health in Toronto, Canada. Her program of research focuses on suicide and suicide prevention, specifically the intersection of gender, culture, and suicide discourses and the experiences related to suicidal thoughts or behaviors of people, families, and clinicians in health care settings. She has also conducted responsive qualitative program evaluations and qualitative research in medical education. She has published in several major journals including *The Lancet Psychiatry*, *Canadian Medical Association Journal* (CMAJ), *Schizophrenia Research*, and the *Canadian Journal of Psychiatry*.

Abbreviations

CBT	Cognitive behavior therapy
HS	Hora somni (medication given at bedtime)
IM	Intramuscular (injection)
MHP	Mental health professional
MSE	Mental status exam
NRT	Nicotine replacement therapy
NSG	Nurse or nursing
PO	Per os (by mouth)
PRN	Pro re nata (taken as needed)
PTSD	Posttraumatic stress disorder
SNRI	Serotonin and norepinephrine reuptake inhibitors
STAT	Statim, meaning immediately
TCA	Tricyclic antidepressants

List of Tables

1

Introduction: Psychiatric Documentation, Power, and Violence

Andrea Daley and Merrick D. Pilling

> Down the other end of the long polished corridor, others recorded their own version of my distress. The files were all about me but I couldn't see me in them.
>
> What I didn't know until I read these notes, is how little regard they had for me as a human being in a desperate existential struggle.
>
> —Mary O'Hagan, *Madness Made Me*

Psychiatric documentation of people's lives is a form of institutionalized power and violence that is usually unnoticed and unquestioned. Historical and contemporary scholarship and grassroots activists have

A. Daley (✉)
School of Social Work, Renison University College, University of Waterloo, Waterloo, ON, Canada
e-mail: andrea.daley@uwaterloo.ca

M. D. Pilling
Women's and Gender Studies in the School of Social Work, University of Windsor, Windsor, ON, Canada
e-mail: Merrick.Pilling@uwindsor.ca

© The Author(s), under exclusive license to Springer Nature Switzerland AG 2021

A. Daley and M. D. Pilling (eds.), *Interrogating Psychiatric Narratives of Madness*, https://doi.org/10.1007/978-3-030-83692-4_1

interrogated psychiatry, psychiatric discourse, and psychiatric institutions for their complicity in violence through their subtle and aggressive imposition of diagnosis, classification, and "treatment." Critical feminist, critical race, decolonial, and Mad scholars and activists have challenged the regulatory and disciplinary role of psychiatry, psychiatric discourse, psychiatric institutions, and associated practices, exposing them as reflecting and reinforcing gendered, raced, sexualized, classed, and sanist social moralities and norms that prop up colonial and white supremacist systems of power (see Abdillahi et al., 2017; Burstow et al., 2014; Daley et al., 2012, 2019; Daley & Ross, 2018; Fanon, 2008; Fernando, 2017; Gorman, 2013; Joseph, 2014; Kanani, 2011; Kilty & Dej, 2018; LeFrançois et al., 2013; Metzl, 2010; Mills, 2017; Pilling et al., 2018; Tam, 2013). While less attended to through such critical analyses, the practice of psychiatric chart documentation is, indeed, central to questions of psychiatric power and violence.

Discipline-specific (e.g., anthropology, medical theory, psychiatry, nursing) interrogations of the medical chart have largely espoused a social constructionist perspective to denaturalize it as a socially produced artifact that reveals the hermeneutic underpinnings of medicine generally, and psychiatry specifically (Leder, 1990; Poirier & Brauner, 1990). Earlier explorations of the interpretive nature of medicine have identified the power of the medical chart to establish the authority of physicians to narrate patient experience, reify relationships of power and rationality, and persuade readers of correct diagnoses (Barrett, 1988; Leder, 1990; Poirier & Brauner, 1990). Similarly, the psychiatric chart has been examined for its discursive representation of psychiatric patients (Coker, 2003),[1] to reveal professional assumptions about patients and their "problems" (Mohr, 1999), to study the transformative effect on patients as an outcome of being "written up" (Barrett, 1988), to interrogate the accuracy of chart documentation (Galasiński & Ziółkowska, 2013), and to question the influence of "patient characteristics" on documentation by mental health professionals (MHPs) (Cradock et al., 2001). Within this interpretative frame, the specific function of language in the psychiatric chart has been explored, albeit minimally and less than critically, to suggest that the psychiatric chart may reflect MHP biases (Martin & Ricciardelli, 2020; Metzl et al., 2016), more often speaking to "societal presuppositions than to the condition of the patient" (Mohr, 1999, p. 1052).[2]

The psychiatric chart has been characterized as a "selective reformulation of verbal accounts and previous records" (p. 59) that serves to "construct a unified story of the patient" (Georgaca, 2013, p. 59). In these reformulations, biological/objective (e.g., symptom identification and medical interventions) and social/subjective (e.g., personal and social context) accounts of distress compete with one another, as multidisciplinary knowledge bases and philosophical commitments (e.g., psychiatry, nursing, social work, occupational therapy, pharmacy) constitute the substance of the psychiatric chart (Georgaca, 2013; Hamilton & Manias, 2006). While each discipline uses language in slightly different ways with varying degrees of frequency and for slightly different ends, documentation typically happens in similarly prescribed ways, relying on a shared bio-psychiatric language (Poirier & Brauner, 1990). Epistemic violence is enacted against patients through institutionalized documentation practices by "objectifying clients through systematic deletion of their perspectives, obscuring the professional's participation and fragmenting the client's experience in order to fit into the standardized sections of a psychiatric report" (Georgaca, 2013, p. 57; see Barrett, 1988). The obscurity of "professional's participation" in documentation is a particularly successful outcome of disciplinary documentation practices that encourage an objective professional gaze. MHPs recede or disappear into the psychiatric chart as it emerges as a "non-storied" text (Lewis, 2017, p. 305), or rather, a neutral and disinterested narrative of distress.

While these examinations offer important insight into the psychiatric chart, their focus on the perspectives, values, and documentation practices of individual, multidisciplinary MHPs fail to critically engage with the ways in which institutional ideology shapes "what people see and say" (Marecek & Gavey, 2013, p. 7) in the context of historically specific interpersonal, social, and institutional processes. This collection of writings challenges the perception of the psychiatric chart as a neutral and disinterested text produced by individual actors and disciplinary documentation practice by shifting the focus of analysis to the power of institutional ideology. The chapters included in this book coalesce to reveal the psychiatric chart as a text that is, in fact, "storied" by institutional ideology that reflects, reinforces, reinterprets, and, at times, resists normative gendered, raced, sexualized, and classed norms, values, and presuppositions.

The analyses draw upon research findings from a research project titled *Cultural Representations of Gender in Psychiatric Narratives*.[3] The overarching goal of the project was to reveal how psychiatry and the psychiatric institution are implicated in the organization of gender relations at the intersection of race, Indigeneity, sexuality, and class. It sought to represent particular standpoints in an effort to investigate how psychiatric discourses participate in the (re)production and sanctioning of gendered ways of being distressed. The project was guided by an earlier pilot study that examined how MHPs' gendered, racialized, sexualized, and classed concepts and images inform their understanding or meaning-making of women's distress (see Daley et al., 2012).[4]

Using inter-related data from the projects cited above, this book has four key aims. First, the book offers a case study, of sorts, to illustrate the ways in which the psychiatric institution implements colonial and white supremacist ideologies. Psychiatry's relationship with colonial and white supremacist ideology is well established through a critical body of literature in which the development of psychiatric theories and practices in the context of the politics of colonialism and white supremacy globally has been explored (Fanon, 2008; Kanani, 2011; Mahone & Vaughn, 2007; Mills, 2017; Morrow & Halinka Malcoe, 2017; Tam, 2013). Colonial psychiatry has been cited as a technology of colonialism, offering colonizers scientific language and mechanisms to, for example, administratively manage people, designate the "normal" mind of the colonized as "abnormal," and relatedly reconfigure resistance to colonization as pathology (i.e., pacifying resistance) (Fernando, 2010; Mahone & Vaughn, 2007; Mills, 2017). In this regard, Mills (2017) articulates precisely the relationship between psychiatry and colonization:

> Psychiatry has been constituted through colonialism and so is always a colonial practice[;] psychiatry and colonialism (even when seemingly operating apart from one another) use similar tools, tools that constitute

> and are built upon the interlacing categories of madness and savagery or primitivity. (p. 130)

Following this, psychiatry is implicated in colonization through the often violent[5] imposition of Western notions of normality rooted in normative gender, race, sexuality, and class, or rather, whiteness. Of course, other psy professions that participate in psychiatric chart documentation are also implicated in colonizing practices and processes in health/mental health care service systems. Nursing and social work scholars, for example, have interrogated how white privilege and racism in their respective professions sustain colonizing thinking and action (Canadian Association for Social Work Education, 2017; Czyzewski & Tester, 2014; Lee & Ferrer, 2014; Mcgibbon et al., 2013; Valderama-Wallace & Apesoa-Varano 2020; Waiter & Nardi, 2019). As such, the chapters in this book center the psychiatric chart to elucidate the ways in which it mediates and organizes the social practices of knowing (LaFrance, 2018) of MHPs beyond psychiatrists to reflect and reproduce the colonial and white supremacist ideologies of the psychiatric institution. Thereby, the aim of the book is to deconstruct the psychiatric chart as a text that serves the administrative interests of the psychiatric institution to manage order and disorder along gendered, racialized, sexualized, and classed lines by concerting the actions of MPHs in and across time and space.

Second, this book seeks to empirically ground psychiatric critical feminist, critical race, decolonial, consumer/survivor, and Mad theorizing of psychiatric oppression and violence at the intersection of distress and gender, race, Indigeneity, sexuality, and class. To this end, the book prioritizes intersectional analyses (Combahee River Collective,1997; Crenshaw, 1991) to consider the ways in which gender normativity, whiteness, and colonization, among other social structures, shape psychiatric narratives of distress. In particular, intersectional analyses highlight the nuanced ways in which colonial and white supremacist ideology is activated in chart documentation to produce qualitatively specific psychiatric narratives of distress and related responses in the psychiatric institution. It is our intention that this collection expands practical knowledge for contemporary consumer/survivor/mad social movements and related community-based activist groups that seek to challenge psychiatric violence and oppression at the intersection of sexism, cissexism, racism, heterosexism, classism, and sanism.

Third, this book will serve as a much needed critically oriented academic and practice resource for health/mental health-related professionals and related education and training programs with respect to clinical documentation practices that meaningfully take into account the social and structural materiality of people's lives and its impact on experiences of distress. It is an urgently needed intervention for ongoing health/mental health professional education and training programs and codes of ethics that continue to approach documentation as a technical exercise rather than a situated process of knowledge production. This collection of writing provides a critical reading of psychiatric documentation, making critical feminist, critical race, decolonial, and Mad Studies relevant to understandings of distress. Thus, it challenges health/mental health education and training programs that socialize MHPs to be "neutral" and disinterested witnesses and recorders of distress (Blair & Smith, 2012; Myklebust et al., 2018; Savaya, 2010).

A final aim of this book is to offer an innovative, critical research methodology for textual analysis, particularly with respect to the critical analysis of cultural norms, values, and discourses as they are reflected in and reproduced in psychiatric documentation practices and institutional ideology. As such, this book seeks to contribute to critical research methodology that is social justice-oriented and transformative in intention and outcome. To this end, given the projects' orientation to institutional ethnography (see below), we note that while data were collected from only one psychiatric institution, the analyses and insights offered in this book are applicable across psy sites. As Smith (2005) points out, "institutions" refer to "complexes embedded in the ruling relations that are organized around a distinctive function, such as education, healthcare, and so on" (p. 225). The ruling relations organized around the function of the psychiatric institution selected for the purpose of this critical exploration are replicated in psychiatric institutions across time and space. As such, the critical methodology underlying this collection of writings underscores that patients' experiences of being "held" to colonial and white supremacist ideologies, including gendered, raced, sexualized, and classed social relations, are similarly meditated and organized by the psychiatric chart beyond this particular institutional setting.

Methods

The projects upon which this book is based used theories and methods informed by institutional ethnography (Smith, 2005), which uses interviews, observations, and texts as data to explore how experiences are bound up in "social relations" (Campbell & Gregor, 2004). Social relations refer to practices and activities that are coordinated and concerted by people's interactions with each other and with shared languages, taking "experience" beyond one's own motivations and intentions. Social relations include textually mediated practices in which texts have the power to "hold people to acting in particular ways" (Campbell & Gregor, 2004, p. 173). Institutional ethnography approaches challenge the idea that constructs, such as gender, race, sexuality, and class, can be abstracted from the practices of people acting within organizations and through texts. Instead, such constructs should be investigated as they emerge within a particular context and through people's activities (DeVault & McCoy, 2004). Using an institutional ethnography lens, the contributors to this book have sought to reveal connections between the psychiatric inpatient charts, activities, and texts that precede them (e.g., organizational policies, everyday practices, dominant discourses on gender, race, sexuality, and class), and activities and texts that follow them (e.g., further charts, patient discharges, etc.). We consider the psychiatric chart as integral to human relations and lived experiences within the psychiatric institution, even when not all forms of data are available (as in the patient's own account of their experiences or exact transcripts of practitioner-patient interviews). Therefore, we view the psychiatric chart as both a product and a producer of people's activities with material, everyday implications.

Research Design

The project involved a large psychiatric institution in Ontario, Canada, and included two phases. In Phase 1, a retrospective chart review of 161 psychiatric inpatient charts was conducted between August 2015 and September 2016 and in July 2017. Phase 2 included three individual semi-structured interviews with institutional key informant hospital administrators to explore institutional connections, organizational processes, and organization-specific texts such as policies related to safety on the ward (physical and sexual), reporting of incidents, risk flagging, and use of restraints and seclusion that shape psychiatric chart documentation. These interviews were informed by and built on eight individual semi-structured interviews with institutional key informants (psychiatrist, psychologist, nurse, social worker, nurse educator, professional practice lead, unit managers) conducted during the pilot phase of the project (2011). The most recent set of MHP interviews were conducted in November 2017 and the earlier set, between January and March 2011. Notwithstanding the time span of the interviews, they serve as a discursively coherent context for our critical examinations of institutional documentation practices. While the method of chart documentation has changed over this period from semi-electronic to fully electronic, the knowledge base, professional and institutional standards, and tools that guide documentation (e.g., multidisciplinary assessment form, mental status examination) have remained relatively static. That is, while the technology of chart documentation has evolved, its institutionalized medical, legal, and ethical/philosophical underpinnings have not.

Retrospective Chart Review

Charts belonging to patients discharged between January 2013 and September 2016 were selected from anonymized discharge lists from four inpatient programs that provide treatment related to particular diagnoses and/or symptoms: schizophrenia, bipolar disorder, major depression disorder, and generalized anxiety disorder.[6] We selected programs that had a large number of patients with longer stays (to allow for rich and plentiful charting data) as well as a range of diagnoses that have been shown to vary by gender, sexuality, race, and class (Beauboeuf-Lafontant, 2007; Blum & Stracuzzi, 2004; Chesler, 2005; Jimenez, 1997; Metzl, 2010; Ussher, 1997, 2011, 2013).

All charts were reviewed sequentially as they were listed on the discharge list until the target number of 30 charts from each unit was met. The chart review included those components of the electronic psychiatric chart that the research team identified as being most likely to be populated with information most relevant to the purpose and scope of the project (Table 1.1). The chart components from which data were extracted included both drop-down menu and narrative formats (see Chapter 2 for more detail).

Table 1.1 Electronic chart components included in the review by category

Emergency department multi-disciplinary assessment	History and physical documentation	Transfer of an inpatient
Referral from hospital to community	Admission note	Client communication
Collateral information	Progress note	Consultation note
Pre-arrival note	Simplified progress note—inpatient	Patient writing/drawings
Information on Form 1	Pass/off-ward privileges documentation	Close/continuous observation documentation
Triage note	AWOL documentation	Spiritual care assessment
Suicide risk assessment	Team treatment plan	Safety and comfort plan
Miscellaneous	Therapeutic intervention/group progress note	Safety plan for coping with suicidal thoughts
Pharmacy assessment	App by physician for Ps assess Form 1	Global assessment of functioning
Mental status exam	Overall impressions of chart	Transfer/admit of inpatient documentation
Emergency department clinical summary	Emergency department patient summary	

Each chart was read through once in an effort to immerse the researcher in the data, with the primary goal of determining its inclusion eligibility. Chart eligibility was based on: (1) length of stay, with eligible charts indicating an admission period of at least two weeks; and (2) confirmation that team members did not have a conflict of interest with the chart (e.g., relationship with the patient). During the initial review of the chart, documentation content related to gender, race, Indigeneity, sexuality, and class was extracted into an anonymized Word document (e.g., relationship status/history; descriptions of hygiene, dress, skin color, etc.; references to sexual behavior and sexual identity; sexuality-related concerns such as relationships, sex, and contraception; and references to housing status, income/social assistance needs, and education).

During the chart selection process, we also attended to representation across the categories of gender, race, Indigeneity, sexuality, and class by maintaining a demographic table of the charts we selected during the selection process (Table 1.2). We relied on the multidisciplinary assessment form completed during admission to collect patient demographic data on gender, sexual orientation, race, Indigeneity, and socioeconomic status/class. Frequently, conflicting or contradictory demographic information was seen in a psychiatric chart. Demographic information on sexual orientation and race and descriptors of body size, for example, collected through the multidisciplinary form in the emergency department was contradicted by descriptions of patients in MHP progress notes. It is important to note that the contradictions in demographic information suggest that while at times patients may have self-identified in a certain way to MHPs, in some instances MHPs completed the form based on their own interpretation of the patient from their specific positionality. Relatedly, we note that the institution's method of taxonomizing gender does not allow for those who identify as non-binary. This is a serious limitation that indicates structural transphobia. In our review of the charts, it was apparent that MHPs were especially ill-equipped to accurately capture a patient's gender identity (Pilling, 2022). In sum, these sorts of documentation contradictions and silences speak to the violence inherent in quantifying identity in a drop-down format

Table 1.2 Chart demographics

Demographics	Cis women	Trans women/trans feminine	Cis men	Trans men/trans masculine	Non-binary	Total
Race						
White	37	8	58	1	4	108
Black	11	2	9	0	0	22
Indigenous	1	1	2	0	0	4
East Asian	3	0	1	0	0	4
South Asian	3	0	4	0	0	7
Latin American	2	0	2	0	0	4
Middle Eastern	3	0	2	0	0	5
Mixed Heritage	2	0	1	0	0	3
Not listed	2	0	1	0	0	3
Other	1	0	0	0	0	1
Total	**65**	**11**	**80**	**1**	**4**	**161**
Sexual orientation						
Lesbian	6	0	0	0	0	6
Gay	0	0	10	0	0	10
Bisexual	2	0	12	0	1	15
Queer	0	1	0	0	1	2
Asexual	0	0	0	0	1	1
Not listed	8	1	10	1	0	20
Do not know	0	4	5	0	1	10
Prefer not to answer	1	0	3	0	0	4
Other: male to female	0	1	0	0	0	1
Other: female to male	0	0	0	1	0	1
Other: transgender	0	2	0	0	0	2
Total	**17**	**9**	**40**	**1**	**4**	**71**
Income						

(continued)

and institutional categorization practices that are neither neutral nor objective.

Notwithstanding the limitations of the multidisciplinary assessment form for the collection of demographic information, we used our demographic table to identify missing and underrepresented social identity and social location categories, employing the purposive selection of charts

Table 1.2 (continued)

Demographics	Cis women	Trans women/trans feminine	Cis men	Trans men/trans masculine	Non-binary	Total
None	1	0	0	0	0	1
0–14,999	35	7	42	1	1	87
15,000–	4	3	6	0	1	13
29,999	5	0	5	0	0	11
30,000–59,999	5	0	6	0	1	10
Do not know	13	1	16	0	1	30
Not listed	1	0	7	0	0	10
Prefer not to answer						
Total	**64**	**11**	**81**	**1**	**4**	**161**
Number of involuntary admissions	39	7	37	0	0	**83**

from the discharge lists to address gaps in the data. In this regard, we purposively selected an additional 14 charts belonging to trans- and non-binary-identified patients, 15 charts identified as belonging to lesbian-, gay-, bisexual-, and queer-identified patients, and 12 charts identified as belonging to bisexual-identified patients. A total of 161 charts were reviewed.

Fifty-nine percent (59) of the psychiatric charts reviewed belonged to cis women and men who were identified on the multidisciplinary assessment form as White-North American (Canadian, American) or White-European (English, Italian, Portuguese, Russian). Less than 2.0% (3) of cis women and men were identified as Indigenous. Almost 7% (11) and less than 5.5% (9) belonged to cis women and men, respectively, who identified as Black-North American (Canadian, American), Black-Caribbean (Barbadian, Jamaican), or Black-African (Ghanaian, Kenyan, Somali). Less than 2% (3) of cis women and men were documented as mixed heritage (Black-African and White-North American), and about 15% (20) were identified as Asian-East (Chinese, Japanese, Korean), Asian-South East (Malaysian, Filipino, Vietnamese), Latin American (Argentinean, Chilean, Salvadorian), or Middle Eastern (Egyptian, Iranian, Lebanese) cis women and men. About 2.5% of cis women and men (4) did not have a racial identity listed (i.e., "not listed" and "other"). Of the 145 charts belonging to cis women and cis men, 30 (21%) documented sexual orientation as non-heterosexual: lesbian (6), gay (10), and bisexual (14); and eighteen (18) charts recorded sexual orientation as "not listed," five (5) as "do not know," and four (4)"prefer not to answer." Almost seven percent (11) of patients were identified as trans women/transfeminine people, with two identified as Black, one as Indigenous, and eight as white. One patient was identified as a white trans man and four were identified as white, non-binary people. Of the 16 trans and gender non-binary charts, four (4) erroneously recorded sexual orientation as transgender; one (1) bisexual; one (1) asexual; two (2) as queer; five (5) as do not know; two (2) as heterosexual; and one (1) was left blank. Fifty-four percent of the charts indicated that patients were living on an income of less than $15,000, with approximately 30% of charts not having income listed.

The most significant discrepancy across the four units from which the psychiatric charts were selected was the disproportionate number (10%) of Black patients represented in the inpatient unit that treated people with a diagnosis of schizophrenia and related disorders. This compares to 0.8%, 2.5%, and 2.5% of Black patients in the other units focused on shorts stay, mood and anxiety, and older patients, respectively. Given that the selection of charts from each unit followed the same protocol, this difference in representation along racial lines is an important observation that is aligned with critical scholarship that interrogates the over-diagnosis of Black people, particularly Black men, with schizophrenia (Knight et al., 2021). For example, Metzl (2010) contextualizes the high prevalence of schizophrenia diagnoses among Black men in relation to shifts in the gendered and racialized construction of schizophrenia from that of "an illness of white feminine docility to one of black male hostility" (p. xv). His analysis underscores how diagnostic criteria and the "conversations that take place" within psychiatric institutions are shaped by racialized structures.

Selected charts were read a second time during which notations related to gender, race, Indigeneity, sexuality, and class (as described above) were extracted from the chart. De-identified extracted data from each chart were manually transferred into individualized Word documents that served as a "text" for analysis (Daley et al., 2012).[7] Careful consideration was given to ensuring that data were not de-contextualized. For example, each excerpt was identified by the profession of the documenting MHP and the day of the inpatient admission during which the documentation occurred (e.g., Day 23).

Once all relevant data were extracted from the total 161 charts, the following steps were undertaken by two research team members (MP and AD): (1) all the texts were read and a summary for each text was created, (2) each summary was read and a memo was created for each summary to capture key themes across the texts, and (3) MP and AD read each other's memos with the intent to refine the key themes. Key themes were presented to the other team members (MG, LR, and JZ) for feedback several times throughout this process. In this regard, critical discourse analysis was used to highlight the social and organizational categories of gender, sexuality, race, and class that already existed in the charts and that functioned to shape chart documentation. Our readings explored ideological commitments underlying psychiatric charts including gendered, racialized, sexualized, and classed assumptions, meanings, and values (Lupton, 1992; Tupper, 2008).

Interviews with Mental Health Professionals

While retrospective chart review studies have provided important insights into chart documentation practices and practitioner attitudes, as described above, several limitations in documentation reveal the weakness of this methodology: incomplete documentation, information that is unrecoverable or unrecorded, difficulty interpreting information found in the charts (e.g., acronyms), and variance across MHPs and patient charts in the quality of information recorded (Gearing et al., 2006). Similarly, patient discomfort engaging in discussions with MHPs about sensitive topics or reluctance on the part of patients and/or MHPs to have sensitive information documented may contribute to a lack of chart documentation and, therefore, represent a further limitation of retrospective chart review methodology. To acknowledge and explore this overarching limitation, the project also included in-depth, face-to-face, semi-structured interviews with key informant MHPs, as described above, to explore the phenomenon of chart documentation as it related to gender, race, Indigeneity, sexuality, and class.

Key informant MHPs employed by the institution were purposively recruited through existing organizational contacts to achieve diverse representation based on professional and institutional roles. In keeping with institutional ethnography methods, additional key informants were identified as interviews progressed and the researchers learned more about relevant organizational processes and organization-specific texts. The interview guide comprised open-ended questions aimed at exploring ruling relations that organized MHPs charting activities: (1) their experiences and ideas about documenting gender, race, sexuality, and class-related information, (2) where they learned to chart in the way they do (decisions to include/not include information), (3) factors that influenced when and where they documented such information, and (4) their concerns about charting in different ways, whether, when, and why deviations from documentation standards occurred. The interview guide was elaborated upon in an iterative process to incorporate data emerging from the interviews and the analysis of organization-specific texts (DeVault & McCoy, 2004). Relevant organization-specific texts related to chart documentation known to the researchers and identified by key informants were selected for review with the specific purpose of gaining information about other relevant texts that might influence chart documentation practices.

All interviews were digitally audio-recorded, transcribed verbatim by a professional transcriber, and anonymized. Multiple readings of each transcript were performed by two research team members to identify major themes related to the research. A constructivist grounded theory method was used to identify thematic content (Charmaz, 2000, 2006).

Ethics

The ethical considerations underpinning the project as well as the intentions and aims of this book thread back to the philosophical and theoretical considerations that shaped the pilot project referenced above (Daley et al., 2012).[8] At its core, the research that informs the analyses presented in this book is best characterized as emancipatory in intent, engaging a process of "studying up" (Harding & Norberg, 2005) to reveal the power of institutional ideology (McColl et al., 2013; Moosa-Mitha, 2005) to reinscribe oppressive colonial and white supremacist social and structural relations.

At the beginning of this project's life, as with any research, we developed our ethics protocol. We approached ethical considerations guided by institutional research ethics boards, relying on the research principles and practices endorsed by the Tri-Council Policy Statement on Ethical Conduct for Research Involving Humans (2018). To this end, the project received ethics approval from the psychiatric institution from which charts were reviewed and from all participating universities, an informed consent was obtained from each MHP key informant prior to the start of each interview. However, a critical question that emerged was how to access inpatient charts. More specifically, we grappled with the decision of whether to work exclusively through the psychiatric institution or with patients themselves to gain access to psychiatric inpatient charts. Through a series of discussions about benefits versus harms and privacy and confidentiality that were informed by a principled orientation to ethics (Preissle, 2007), we decided to work through the institution. This meant requesting ethics approval for a waiver of informed consent to access patient charts,[9] or rather, reviewing psychiatric charts without having the explicit consent of patients. This was a difficult decision to make in light of our collective commitment to research practices that were informed by social justice principles and research processes and outcomes that promote individual, institutional, and structural transformation. We embraced an approach that recognized the critical importance of scrutinizing and assessing the request for a waiver of informed consent from a patient rights perspective. Central to our discussions was a consideration of the rights of patients to decide about participation/nonparticipation, feasibility issues related to obtaining informed consent, the risk for patients of obtaining informed consent, and the potential benefits of conducting the research. We offer an overview of this process and rationale for the decision below.

The research team began by exploring, with the institution's ethics review board manager, the feasibility of obtaining informed consent from patients. We outlined a process that involved several steps: medical records staff pre-selecting inpatient charts of discharged patients for review, identification of a MHP who had a prior relationship with the patient, medical records staff liaising with the MHP to inform them of the study and informed consent process, the MHP reaching out to the discharged patient to explore participation, and obtaining informed consent, if participation was indicated by the patient. A thorough evaluation of this process yielded several feasibility barriers to the completion of the study: MHP workload concerns, financial remuneration for the MHPs, the pre-selection of charts by medical records staff that did not meet the eligibility requirements, and patients no longer residing at the address listed on the chart. The latter two barriers suggested that a large number of charts would need to be accessed in an effort to meet the targeted sample size, thereby, signaling significant workload for both medical records staff and MHPs.

Beyond the feasibility barriers, we identified a number of potential risks for discharged patients. Central to our decision-making was our concern that seeking informed consent might, inadvertently, compromise people's right to privacy and confidentiality with regard to their medical histories. That is, despite taking measures to maintain privacy and confidentiality by, for example, sending letters to patients' home addresses in nondescript envelopes or leaving telephone messages with minimal information, contacting discharged patients to obtain informed consent might lead to the inadvertent disclosure of their involvement with psychiatric services. The research team was concerned that this would pose a particular risk for economically and socially marginalized people, who often reside in settings that afford little privacy (e.g., shelters, boarding homes). While we were careful not to perpetuate the stigma that is often associated with a mental health diagnosis and involvement with psychiatric and mental health services through this consideration, it is also important to recognize the existence of stigma and its very real consequences on people's lives.

The research team also considered that contacting patients for the purpose of informed consent would inadvertently result in their feeling particularly vulnerable during future contact with psychiatric and mental health services and professionals. More specifically, the team considered the potential risk that people would not seek support and treatment in the future because of feelings of vulnerability that emerged when contacted for research consent purposes. Consequently, it was determined that contacting patients for informed consent purposes could not only inadvertently result in people feeling particularly vulnerable during future contact with MHPs but could reduce the likelihood of their engagement with mental health services and professionals during times of crisis. It was noted that this may be a particular risk for cis and trans women, trans men, and racialized, Indigenous, lesbian, gay, bisexual, queer, and poor people who were already experiencing barriers to mental health services as a result of their membership in communities that are subjected to varying degrees of social marginalization.

The team considered the impact of power in the MHP-patient relationship. While MHPs often work to minimize or ameliorate the power imbalance that exists in their relationships with patients, it is difficult to achieve. Consequently, we were concerned that MHPs may implicitly influence patients' decisions to provide consent, namely patients may desire to make a favorable impression on MHPs or fear that not participating may have a negative impact on their relationship with the MHP and care during future admissions.

Finally, we considered the potential of the research to redress the legacy and ongoing harm of the psychiatric regulation and disciplining of cis and trans women, trans men, and racialized, Indigenous, lesbian, gay, bisexual, and poor people. We critically reflected on the need to weigh the potential harm of working through the psychiatric institution against the potential harm people encounter during their interactions with the psychiatric institution. In the end, we reached consensus that working through the institution posed no harms greater than those potentially encountered by people during their interactions with the institution.

Without doubt, recognition of people's rights to provide free and informed consent in research is critical, particularly when research studies involve people made vulnerable by inequitable and violent social structures. In this regard, it is important to note that this decision will not be uncontested. Likely there are readers of this book who will disagree with our decision around this, and we are called to account for this decision on an ongoing basis. That is, if we are not able to effect individual, institutional, and structural change as a result of this work then the risks to people of having their records reviewed by us without their explicit consent does not outweigh the benefits. It is our sincere intention that this book contributes to such change.

Organization of the Book

The collection of writings offered in this book represents key themes identified through the analysis of 161 psychiatric inpatient charts and 11 key informant MHP interviews. Most of the analyses presented in the book are based solely on the chart review data, with the interview data from the project (November 2017) and the pilot project (January and March 2011) indirectly supporting the analyses. However, one analysis integrates MHP interview data from both sets of interviews in an effort to represent a range of MHP perspectives (see section "Research Design"). Collectively, the chapters offer a critically robust and comprehensive interrogation of psychiatric narratives of distress at the intersection of gender, race, Indigeneity, sexuality, and class by deconstructing psychiatric chart documentation practices.

In Chapter 2, Margaret Gibson explicates how the psychiatric chart achieves categorical differentiation asking, "What shapes the ways in which differences between patients are written about, and what happens as a result?" She examines how what professionals perceive and write is shaped through documentation practices and forms, and how the documents produced, in turn, shape differences in what providers do and patients experience. In this, Gibson reveals how actions and texts are integral to the way differences between patients come to matter, identifying, for example, key documentation elements such as gendered, racialized, and classed notions of "appropriateness." Throughout her analysis, she explores how documentation processes and forms constrain not only the people who are documented and the people who document, but also the larger social imagination. Gibson concludes by pointing to the electronic psychiatric chart as a governing apparatus, working in tandem with long-standing assessment practices within and across psychiatric institutions to implement colonial and white supremacist ideologies. Beyond the substantive analysis, Chapter 2 offers an overview of the psychiatric electronic chart, orienting readers for the analyses presented in subsequent chapters.

Andrea Daley, in Chapter 3, interrogates how gender, sexuality, race, and class structure the normative base of psychiatry and the psychiatric institution by asking what about gender is most relevant in chart documentation. Her analysis draws on two key themes, *gendered aesthetics* and *recognizablegender = credible narrator*, to reveal how the psychiatric institution serves as a social and structural mechanism to protect white hegemonic femininity and masculinity and a normative gender order. Daley concludes by drawing attention to the psychiatric chart as an institutional artifact that deploys white supremacist ideology, valorizing whiteness (white norms, culture, and people), while ordering and disordering patients in a web of hierarchies between women and men as well as among women and among men, as constitutive of normative gender relations.

In Chapter 4, authors Lori Ross and Lucy Costa offer a content analysis of chart documentation for 12 bisexual patients to examine where, when, and how bisexuality was documented in the psychiatric record, as well as where and when it remained invisible. Through an integrated Mad Studies and psy discipline (e.g., psychology, psychiatry) lens, their analysis highlights the ways in which the structure of the psychiatric record, together with the institutional practice of charting and the knowledge and attitudes of individual MHPs, make bisexuality (in)visible in psychiatric admission. They examine how common social constructions of bisexuality are reinforced by the institution of psychiatry, as well as how meaningful attention to bisexuality contributes to the patient- and clinician-defined success of a psychiatric admission. Importantly, Ross and Costa address a gap in Mad Studies and psy disciplines literature, bringing a Mad Studies orientation (i.e., social, political, historical) to meet the individual-level insights that have been drawn from psy disciplines to advance a critical examination of the psychiatrization of bisexual people.

In Chapter 5, Juveria Zaheer assumes a social interactionist orientation to explore the framing of restraint and use of seclusion in psychiatric chart documentation. More specifically, she uncovers the ways in which patients' trauma experiences and the enactment of trauma by MHPs are minimized through the institutionalized mechanisms of framing patients' concerns related to institutional practices and professional decisions as disruptive or demanding; asymmetrical reporting of MHP-patient interactions that minimize the MHP's role in a patient's distress; foregrounding providers' emotional experiences rather those of the patients in chart documentation; and the use of the passive voice in chart documentation to underscore the perceived inevitability of the restraint or seclusion event. Writing from the perspective of an emergency department psychiatrist, Zaheer reflects on her responsibility, and that of the institution, to critically consider and respond to the violence inherent in restraint and seclusion practices.

Merrick Pilling, in Chapter 6, explores how rape culture shapes the ways in which MHPs document sexual violence in the charts of patients diagnosed with psychotic illness. His analysis includes an investigation of the intersection of rape culture with sanism and anti-Black sanism (Meerai et al., 2016). To this end, he explores five themes emerging from the data: (1) *normalization of sexual violence*, (2) *sexual violence reconfigured as delusion*, (3) *pathologizing the impact of sexual violence*, (4) *what about the perpetrators? invisibilizing acts of sexual violence*, and (5) *sexual violence as a symptom of psychosis*. He concludes with a discussion of the institution's responsibility to address sexual violence within and outside of the psychiatric institution.

We conclude the book with Chapter 7, threading together key points of analysis offered by the contributors and raising critical questions to further advance these considerations. We pay particular attention to suggested tactics for approaching psychiatric documentation in ways that avoid collapsing attention into what those designated as patients "do" and "are" and focus on the ways in which the institutional structures and relations that surround each document can be usefully interrogated.

Before concluding this introduction, we qualify this collection of writings by offering important analytical considerations related to the limitations of the retrospective chart review methodology and data collection methods used in the project, documentation patterns of MHPs, and contributors' positionality. First, as stated in the Methods section, there are inherent limitations to a retrospective chart review methodology with respect to the absence and quality of both patient and MHP experiences, perspectives, and actions. These limitations suggest that the psychiatric chart may not fully represent the practices and attitudes of MHPs, but rather is more likely a proxy measure of their responses to particular patient and MHP interactions, discussions, situations, and information. Of critical importance, the analyses presented in this book are premised on understanding that all therapeutically relevant observations, conversations, insights, and actions are documented as directed by the medical and legal structures of the psychiatric institution and professional regulatory bodies. Notwithstanding our position, we recognize that the limitations of the methodology could limit or contradict the inferences made in the analyses, as well as the overall conclusions of the book.

Second, the analyses may be limited in that the chart excerpts are those of various rather than single psychiatrists, nurses, social workers, recreational therapists, and pharmacists. Relatedly, given that all chart excerpts are anonymized (although the analyses do indicate the profession of the writer, e.g., psychiatrist, nurse, social worker), it is difficult to decipher from the chart files whether notations on a patient's chart or across patient charts were made by the same MHP or several. This limits the analysis with respect to comparing an MHP's documentation patterns within and across psychiatric charts.

Third, the analyses presented in this collection of writings are limited by longstanding institutionalized documentation practices, as reflected in the structure of the electronic record software. While institutional documentation practices require MHPs to identify themselves by name and profession, they do not yield information about their social identities (e.g., gender, race, Indigeneity, sexuality, and class). As such, our review of the psychiatric inpatient charts did not allow for the identification of MHPs' social identities and locations. Having this type of information would contribute to a more robust analysis of power dynamics during patient-MHP interactions. However, while it is important to attend to power dynamics within the specificities of a patient-MHP dyad, we also understand power dynamics as shaped by the ways in which patients and MHPs are positioned in relation to colonial and white supremacist ideologies that undergird the psychiatric institution. Thus, we might consider that all MHPs, regardless of gender, race, sexuality, and class represent and operationalize the norms and values of the psychiatric institution.

Fourth, the contributors to this volume are diverse in their personal and professional lived experiences with respect to their relationship to the mental health system and psychiatry, the psychiatric institution, diagnosis and treatment, theoretical orientations, and activist work toward gender, racial, queer and trans, economic, and consumer/survivor/mad justice. So, while we share an unwavering commitment to critical inquires and practices that contribute to deconstructing and transforming systems of power that sustain injustice, our entry points differ. We hope that you "hear" our shared, yet positioned, voices in this collection of writing.

A final note of consideration. The project on which this book rests centers on the psychiatric chart, an institutional text, as the unit of analysis. However, it is imperative that we remember that people are at the heart of this interrogation. Just as the psychiatric chart can function to "disappear" its writers (i.e., MHPs), it is equally forceful in erasing the people that are being written about. As Mary O'Hagan (2015) observed, "the files were all about me but I couldn't see me in them." Time and time again, through the chart review process we witnessed the power and violence of psychiatric discourse and the invocation of colonial and white supremacist ideologies to subvert people's right to narrate their distress as it was produced and expressed at the intersections of gender, race, Indigeneity, sexuality, and class. Our hope is that this critical account of the psychiatric chart serves to rupture this power and violence.

Acknowledgements Sections of this chapter draw on earlier publications by the authors:

Pilling M. D., Daley A., Gibson M. F., Ross L. E., & Zaheer J. (2018). Assessing "insight," determining agency and autonomy: Implicating social identities. In J. Kilty & E.Dej (Eds.), *Containing madness: Gender and "psy" in institutional contexts* (pp. 191–213). Cham: Palgrave Macmillan.

Daley, A., Costa, L., & Ross, L. (2014). Doing critical feminist research: A collaboration. In S. Wahab, B. Anderson-Nathe, & C. Gringeri (Eds.), *Feminism in social work research* (pp. 19–35). New York: Routledge.

Daley, A., Costa, L., & Ross, L. (2012). (W)righting women: Constructions of gender, sexuality and disorder through psychiatric documentation practices. *Culture, Health and Sexuality, 14*(8), 955–969. https://doi.org/10.1080/13691058.2012.712718

The authors extend their gratitude to Opal Sparks and Dawnmarie Harriott for their support of the project, key insights, and contributions to analytical themes. Their rich and critical perspectives on psychiatry, the mental health system, and patient experiences strengthen the book in innumerable ways.

The authors wish to thank Madeline Samsa for administrative assistance in the preparation of this chapter.

The book draws on research supported by the Social Sciences and Humanities Research Council and the Robert Harding & Lois Claxton Humanities and Social Sciences Endowment Fund, University of Waterloo.

Notes

1. We acknowledge that there is extensive discussion and debate surrounding the language that is used to describe those who use mental health services (see Burstow, 2013). In this and subsequent chapters, we deliberately use the term "patient," given that (1) all of the individuals whose charts we reviewed were psychiatric inpatients, some of them admitted involuntarily, and thus terms that imply more agency (e.g., consumer, client) may not be applicable and (2) that our only data source was chart records, the voices of these individuals are not represented; we are only speaking on the basis of the medical providers who documented their admissions. Thus, we do not know what identity-based terms (e.g., Mad person) they might have preferred for themselves. In this context, it seems most appropriate to use the language of "patient," which we believe best reflects the power relationships inherent in the chart documentation. In short, we employ the term to indicate how the people whose charts we analyzed were interpellated by the psychiatric institution. In referring to specific patients' chart records, however, we the authors apply anonymized pseudonyms to facilitate discussion of the findings. We apply the same pseudonym to each chart across

chapters. For example, the pseudonym Cindi is used in Chapters 2 and 3 to refer to the same chart/patient.
2. For a more detailed account of this body of literature, see Daley et al. (2012).
3. The research team for the project *Cultural Representations of Gender in Psychiatric Narratives* included Andrea Daley, Merrick Pilling, Margaret Gibson, Lori E. Ross, and Juveria Zaheer. Key preliminary findings were shared with two community members, Opal Sparks and Dawnmarie Harriott, for their insight and feedback.
4. The research team for the pilot study included Andrea Daley, Lucy Costa, and Lori E. Ross.
5. We note that violence takes many forms such as verbal, physical, emotional, sexual, and epistemic. This may be particularly true for mental health service users/survivors who are subjected to sanist attitudes and behaviors as a result of living with distress experiences and having received a diagnosis of mental disorder. Some individuals experience psychiatric diagnosis and treatment as violent, as they are held to sanist and normative expressions of distress.
6. While we structured the chart selection and review in relation to these inpatient programs and associated diagnoses, we recognize that individuals often receive multiple and/or inaccurate diagnoses that are not fully aligned with the programs to which they are admitted. As such, there are more than the four identified diagnoses shaping inpatients' experiences related to psychiatric documentation, and, therefore, the analyses are not organized around particular diagnoses.
7. To maintain anonymity and confidentiality, no item of information that would enable the identification of any patient/client or MHP was extracted from the charts. We did not keep a data-linking list of any sort. Thus it is not possible to link data to a particular patient/client or MHP. In an effort to enhance the rigor and comprehensiveness of our analysis, we maintained detailed field notes on the chart selection process. The field notes carefully detail the rationale for each chart's inclusion or exclusion for the purpose of conducting an analysis to identify trends across the charts that were not selected, and to explore what such trends say about psychiatric discourses as described above.
8. The discussion and decision-making process regarding the request for a waiver of informed consent occurred in the pilot study phase of the project among the research team members. The same ethical considerations, protocols, and procedures were adopted in the *Cultural Representations of*

Gender in Psychiatric Narratives project. For a full discussion of the ethical considerations engaged during the pilot phase, see Daley et al. (2014).

9. In the Canadian province in which the research was conducted, a Patient Health Information Protection Act (PHIPA) research plan is required from any investigator who is proposing to conduct research without explicit patient consent. The PHIPA research plan for the project was submitted to, and approved by, the research ethics board of the psychiatric hospital from which inpatient charts were reviewed. The plan articulated that only the principal investigator and research coordinator would have access to inpatient psychiatric charts and, therefore, nonanonymized personal health information, and that no item of information that would enable the identification of any patient/client and/or mental health service provider would be extracted from the charts (e.g., patients'/clients' and MHPs' names, telephone numbers, and addresses).

References

Abdillahi, I., Meerai, S., & Poole, J. (2017). When the suffering is compounded: Towards anti-Black sanism. In S. Wehbi & H. Parada (Eds.), *Re-imagining anti-oppression social work practice* (pp. 87–98). Canadian Scholars Press.

Barrett, T. J. (1988). Clinical writing and the documentary construction of schizophrenia. *Culture, Medicine and Psychiatry, 12*(3), 265–299. https://doi.org/10.1007/BF00051971

Beauboeuf-Lafontant, T. (2007). "You have to show strength": An exploration of gender, race and depression. *Gender & Society, 21*(1), 28–51.

Blair, W., & Smith, B. (2012). Nursing documentation: Frameworks and barriers. *Contemporary Nurse, 41*(2), 160–168. https://doi.org/10.5172/conu.2012.41.2.160

Blum, L. M., & Stracuzzi, N. F. (2004). Gender in the Prozac nation: Popular discourse and productive femininity. *Gender & Society, 18*(3), 269–286. https://doi.org/10.1177/0891243204263108

Burstow, B. (2013). A rose by any other name: Naming and the battle against psychiatry. In B. A. Le Francois, R. Menzies, & G. Reaume (Eds.), *Mad matters: A critical reader in Canadian Mad Studies* (pp. 79–90). Canadian Scholars' Press. https://doi.org/10.1111/inm.12038

Burstow, B., LeFrançois, B., & Diamond, S. (2014). *Psychiatry disrupted: Theorizing resistance and crafting the revolution.* McGill-Queen's University Press.
Campbell, M., & Gregor, F. (2004). Theory "in" everyday life. In W. K. Carroll (Ed.), *Critical strategies for social research* (pp. 170–180). Canadian Scholars' Press.
Canadian Association for Social Work Education. (2017). *Statement of complicity and commitment to change.* https://caswe-acfts.ca/about-us/our-commitment-to-change/. Accessed 14 December 2020.
Canadian Institutes of Health Research, Natural Sciences and Engineering Research Council of Canada, and Social Sciences and Humanities Research Council. (2018). *Tri-Council policy statement: Ethical conduct for research involving humans.* https://ethics.gc.ca/eng/documents/tcps2-2018-en-interactive-final.pdf. Accessed 28 January 2021.
Charmaz, K. (2000). Grounded theory: Objectivist and constructivist methods. In N. K. Denzin & Y. S. Lincoln (Eds.), *Handbook of qualitative research* (2nd ed., pp. 509–535). Sage.
Charmaz, K. (2006). *Constructing grounded theory: A practical guide through qualitative analysis.* Sage.
Chesler, P. (2005). *Women and madness.* Doubleday.
Coker, E., M. (2003). Narrative strategies in medical discourse: Constructing the psychiatric "case" in a non-Western setting. *Social Science & Medicine, 5*(5), 905–916. https://doi.org/10.1016/S0277-9536(02)00459-8
Combahee River Collective. (1997). *Combahee River Collective statement.* https://www.blackpast.org/african-american-history/combahee-river-collective-statement-1977/. Accessed 3 November 2020.
Cradock, J., Young, A. S., & Sullivan, G. (2001). The accuracy of medical record documentation in schizophrenia. *Journal of Behavioral Health Services & Research, 28*(4), 456–465. https://doi.org/10.1007/BF02287775
Crenshaw, K. (1991). Mapping the margins: Intersectionality, identity politics, and violence against women of color. *Stanford Law Review, 43*, 1241–1299. https://doi.org/10.2307/1229039
Czyzewski, K., & Tester, F. (2014). Social work, colonial history and engaging Indigenous self-determination. *Canadian Social Work Review, 31*(2), 211–226. https://doi.org/10.1080/09503153.2018.1483493
Daley, A., Costa, L., & Beresford, P. (Eds.). (2019). *Madness, violence, and power: A critical collection.* University of Toronto Press.

Daley, A., Costa, L., & Ross, L. (2012). (W)righting women: Constructions of gender, sexuality and disorder through psychiatric documentation practices. *Culture, Health and Sexuality, 14*(8), 955–969. https://doi.org/10.1080/13691058.2012.712718

Daley, A., Costa, L., & Ross, L. (2014). Doing critical feminist research: A collaboration. In S. Wahab, B. Anderson-Nathe, & C. Gringeri (Eds.), *Feminism in social work research* (pp. 19–35). Routledge.

Daley, A., & Ross, L. E. (2018). Uncovering the heteronormative order of the psychiatric institution: A queer reading of chart documentation and language use. In J. Kilty & E. Dej (Eds.), *Containing madness: Gender and "psy" in institutional contexts* (pp. 169–189). Palgrave Macmillan.

DeVault, M. L., & McCoy, L. (2004). Institutional ethnography: Using interviews to investigate ruling relations. In W. K. Carroll (Ed.), *Critical strategies for social research* (pp. 191–205). Canadian Scholars' Press.

Fanon, F. (2008). *Black skin, white mask*. Grove Press.

Fernando, S. (2010). *Mental health, race and culture*. Palgrave.

Fernando, S. (2017). *Institutional racism in psychiatry and clinical psychology: Race matters in mental health*. Palgrave Macmillan.

Galasiński, D., & Ziółkowska, J. (2013). Managing information: Misrepresentation in the patient's notes. *Qualitative Inquiry, 19*(8), 589–599. https://doi.org/10.1177/1077800413494346

Gearing, R. E., Mian, I. A., Barber, J., & Ickowicz, A. (2006). A methodology for conducting retrospective chart review research in child and adolescent psychiatry. *Journal of the Canadian Academy of Child and Adolescent Psychiatry, 15*(3), 126–134.

Georgaca, E. (2013). Social constructionist contributions to critiques of psychiatric diagnosis and classification. *Feminism & Psychology, 23*(1), 56–62. https://doi.org/10.1177/0959353512467967

Gorman, R. (2013). Thinking through race, class, and mad identity politics. In B. A. LeFrançois, R. Menzies, & G. Reaume (Eds.), *Mad matters: A critical reader in Canadian Mad Studies* (pp. 269–280). Canadian Scholars' Press.

Hamilton, B., & Manias, E. (2006). "She's manipulative and he's right off": A critical analysis of psychiatric nurses' oral and written language in the acute inpatient setting. *International Journal of Mental Health Nursing, 15*(2), 84–92. https://doi.org/10.1111/j.1447-0349.2006.00407.x

Harding, S., & Norberg, K. (2005). New feminist approaches to social science methodologies: An introduction. *Signs, 30*(4), 2009–2015. https://doi.org/10.1086/428420

Jimenez, M. (1997). Gender and psychiatry: Psychiatric conceptions of mental disorders in women, 1960–1994. *Affilia, 12*(2), 154–175.
Joseph, A. (2014). A prescription for violence: The legacy of colonization in contemporary forensic mental health and the production of difference. *Critical Criminology, 22*, 273–292. https://doi.org/10.1007/s10612-013-9208-1
Kanani, N. (2011). Race and madness: Locating the experiences of racialized people with psychiatric histories in Canada and the United States. *Critical Disability Discourse, 3*, 1–4. https://cdd.journals.yorku.ca/index.php/cdd/article/view/31564/31232
Kilty, J., & Dej, E. (Eds.). (2018). *Containing madness: Gender and "psy" in institutional contexts*. Palgrave Macmillan.
Knight, S., Jarvis, E. G., Andrew, G., Ryder, A. G., Lashley, M., & Rousseau, C. (2021). Ethnoracial differences in coercive referral and intervention among patients with first-episode psychosis. *Psychiatric Services*. https://doi.org/10.1176/appi.ps.202000715.
LaFrance, M. (2018). Institutional ethnography. In P. Liamputtong (Ed.), *Handbook of research methods in health social sciences* (pp. 457–468). Springer.
Leder, D. (1990). Clinical interpretation: The hermeneutics of medicine. *Theoretical Medicine, 11*(1), 9–24. https://doi.org/10.1007/BF00489234
Lee, E. O. J., & Ferrer, I. (2014). Examining social work as a Canadian settler colonial project: Colonial continuities of circles of reform, civilization, and in/visibility. *CAOS: The Journal of Critical Anti-Oppressive Social Inquiry, 1*, e1234p.
LeFrançois, B. A., Menzies, R., & Reaume, G. (Eds.). (2013). *Mad matters: A critical reader in Canadian Mad Studies*. Canadian Scholars' Press.
Lewis, B. (2017). Depression memoirs in the circuits of culture: Sexism, canism, neoliberalism, and narrative identity. *Philosophy, Psychiatry, & Psychology, 24*(4), 303–306. https://doi.org/10.1353/ppp.2017.0043
Lupton, D. (1992). Discourse analysis: A new methodology for understanding the ideologies of health and illness. *Australian Journal of Public Health, 16*(2), 145–150. https://doi.org/10.1111/j.1753-6405.1992.tb00043.x
Mahone, S., & Vaughn, M. (Eds.). (2007). *Psychiatry and empire*. Palgrave Macmillan.
Marecek, J., & Gavey, N. (2013). DSM-5 and beyond: A critical feminist engagement with psychodiagnosis. *Feminism & Psychology, 23*(1), 3–9. https://doi.org/10.1177/0959353512467962

Martin, K., & Ricciardelli, R. (2020). How forensic mental health nurses' perspectives of their patients can bias healthcare: A qualitative review of nursing documentation. *Journal of Clinical Nursing, 29*, 2482–2494. https://doi.org/10.1111/jocn.15264

McColl, M. A., Adair, B., Davey, S., & Kates, N. (2013). The learning collaborative: An approach to emancipatory research in disability studies. *Canadian Journal of Disability Studies, 2*(1), 71–93. https://doi.org/10.15353/cjds.v2i1.71

Mcgibbon, E., Mulaudzi, F., Didham, P., Barton, S., & Sochan, A. (2013). Toward decolonizing nursing: The colonization of nursing and strategies for increasing the counter-narrative. *Nursing Inquiry, 21*(3), 179–191. https://doi.org/10.1111/nin.12042

Meerai, S., Abdillahi, I., & Poole, J. (2016). An introduction to anti-Black sanism. *Intersectionalities: A Global Journal of Social Work Analysis, Research, Polity, and Practice, 5*(3), 18–35.

Metzl, J. M. (2010). *The protest psychosis: How schizophrenia became a Black disease*. Beacon Press.

Metzl, J. M., McClelland, S. I., & Bergner, E. (2016). Conflations of marital status and sanity: Implicit heterosexist bias in psychiatric diagnosis in physician-dictated charts at a Midwestern medical center. *Yale Journal of Biology and Medicine, 89*(2), 247–254.

Mills, C. (2017). Global psychiatrization and psychic colonization: The coloniality of global mental health. In M. Morrow & L. Halinka Malcoe (Eds.), *Critical inquiries for social justice in mental health* (pp. 114–137). University of Toronto Press.

Mohr, W. K. (1999). Deconstructing the language of psychiatric hospitalization. *Journal of Advanced Nursing, 29*(5), 1052–1059. https://doi.org/10.1046/j.1365-2648.1999.00997.x

Moosa-Mitha, M. (2005). Situating anti-oppressive theories within critical and difference-centered perspectives. In L. Brown & S. Strega (Eds.), *Research as resistance: Critical, Indigenous, and anti-oppressive approaches* (pp. 38–72). Canadian Scholars' Press.

Morrow, M., & Halinka Malcoe, L. (Eds.). (2017). *Critical inquiries for social justice in mental health*. University of Toronto Press.

Myklebust, K. K., Bjørkly, S., & Råheim, M. (2018). Nursing documentation in inpatient psychiatry: The relevance of nurse-patient interactions in progress notes—A focus group study with mental health staff. *Journal of Clinical Nursing, 27* (3-4), e611–e622. https://doi.org/10.1111/jocn.14108

O'Hagan, M. (2015). *Madness made me* [Video]. https://vimeo.com/132294911. Accessed 4 January 2021.
Pilling, M. D. (2022). *Queer and trans madness: Struggles for social justice.* Palgrave Macmillan.
Pilling, M. D., Daley, A., Gibson, M. F., Ross, L. E., & Zaheer, J. (2018). Assessing "insight", determining agency and autonomy: Implicating social identities. In J. Kilty & E. Dej (Eds.), *Containing madness: Gender and "psy" in institutional contexts* (pp. 191–213). Palgrave Macmillan.
Poirier, S., & Brauner, D. J. (1990). The voices of the medical record. *Theoretical Medicine, 11*(1), 29–39. https://doi.org/10.1007/BF00489236
Preissle, J. (2007). Feminist research ethics. In S. N. Hesse-Biber (Ed.), *The handbook of feminist research: Theory and praxis* (pp. 515–532). Sage.
Savaya, R. (2010). Enhancing student awareness of the importance of full and accurate documentation in social work practice. *Social Work Education, 29*(6), 660–669. https://doi.org/10.1080/02615470903552006
Smith, D. E. (2005). *Institutional ethnography: A sociology for people.* AltaMira Press.
Tam, L. (2013). Whither indigenizing the mad movement? Theorizing the social relations of race and madness through conviviality. In B. A. LeFrançois, R. Menzies, & G. Reaume (Eds.), *Mad matters: A critical reader in Canadian Mad Studies* (pp. 281–297). Canadian Scholars' Press.
Tupper, K. W. (2008). Drugs, discourses and education: A critical discourse analysis of a high school drug education text. *Discourse: Studies in the Cultural Politics of Education 29*(2), 223–238. https://doi.org/10.1080/01596300801966864
Ussher, J. M. (1997). Framing the sexual "other": The regulation of lesbian and gay sexuality. In *Body talk: The material and discursive regulation of sexuality, madness and reproduction* (pp. 131–58). Routledge.
Ussher, J. M. (2011). *The madness of women: Myth and experience.* Routledge.
Ussher, J. M. (2013). Diagnosing difficult women and pathologising femininity: Gender bias in psychiatric nosology. *Feminism & Psychology, 23*(1), 63–69. https://doi.org/10.1177/0959353512467968
Valderama-Wallace, C. P., & Apesoa-Varano, E. C. (2020). "The problem of the color line": Faculty approaches to teaching social justice in baccalaureate nursing programs. *Nursing Inquiry, 27*(3), e12349. https://doi.org/10.1111/nin.12349
Waiter, R., & Nardi, D. (2019). Nursing colonialism in America: Implications for nursing leadership. *Journal of Professional Nursing, 35*(1), 18–25. https://doi.org/10.1016/j.profnurs.2017.12.013

2

Forming the Chart: Texts, Actions, and Differences

Margaret F. Gibson

Introduction

Charting is an institutional practice that conceals almost as much as it reveals. Mental health service professionals (MHPs) interact with the forms and texts of psychiatric charts throughout their day and are ever mindful that documentation is a core requirement of what they do as professionals and employees. They are accountable to policies, forms, and professional codes of ethics. The people who are assessed at, admitted to, discharged from, or treated through psychiatric institutions are aware that there are "records" of their interactions and that these documents cross time and space to link what happened "then" with what is happening "now" and what could happen "next." Interconnected institutional systems as diverse as professional regulators, child welfare

M. F. Gibson (✉)
School of Social Work, Renison University College, University of Waterloo, Waterloo, ON, Canada
e-mail: Margaret.gibson@uwaterloo.ca

© The Author(s), under exclusive license to Springer Nature Switzerland AG 2021

A. Daley and M. D. Pilling (eds.), *Interrogating Psychiatric Narratives of Madness*,
https://doi.org/10.1007/978-3-030-83692-4_2

systems, courts, and income support programs can request forms and assessments based on psychiatric charts. In many ways, the words that are written in and through psychiatric encounters become their most durable elements.

However, when we look at an institution as a physical space with people going in and out of its doors, we don't usually see this continual flow of words and documents: diagnoses, forms, signatures, credentials, descriptions, justifications, explanations, and regulations jumbled together in what looks like bland and unassuming prose. We may regard records as transparent and reliable reflections of what happened, necessary elements rather than core drivers of how people act, interact, and react. The instructions, conversations, decisions, emotions, changes, omissions, relationships, inconsistencies, and interpretations behind the documents are so readily obscured.

In contrast, this chapter starts from a different understanding of psychiatric records. It examines how the things that professionals perceive and write are shaped through documentation requirements, regulations, and forms, and how the texts produced, in turn, shape differences in what providers do and patients experience. In other words, this analysis investigates the sequencing of actions and texts as an integral way in which differences between people come to matter (Gibson, 2016). Investigating the "how" of charting difference makes this crucial context around and through the charts assailable to researchers and advocates. It explores how the apparatus and practice of documentation affect the everyday "doings" of the psychiatric institution and how documentation can become a core contributor to inequitable encounters and experiences (Smith, 1999, 2005). The central question this chapter addresses is, "what shapes the ways in which differences between patients are written about, and what happens as a result?" In response, this chapter identifies key elements that reproduce raced/gendered/classed/able-bodied difference, such as electronic charting practices and designations of "appropriateness." Throughout, this analysis explores how documentation processes and forms constrain not only the people who are documented and the people who document, but also our larger social imagination.

Background

Institutional power resides, to a great extent, in who charts and who is charted. People document and people are documented in every psychiatric institution, and documentation expands how people can affect each other across time and space. Words wander beyond moments and walls to produce ideas, dynamics, and possibilities between people and systems, while seldom calling direct attention to themselves. Every action in an institutional context occurs in relation to such texts, but the practices of charting are almost never seen as worth examining by institutions, regulators, and providers beyond "did it get charted?" Professionals are taught "if it isn't in the chart, it didn't happen" but are seldom asked to examine what the process of reading, writing, and reproducing charts actually does to their relations, actions, and decisions (Flanagan et al., 2009). Patients learn to be wary of charts but also to know that this is a domain largely out of their control.

Differences matter in and through charting practices. In spite of the standardized policies and forms that shape intakes, discharges, referrals, diagnoses, and assessments, documentation cannot be seen as "one size fits all." The differences in "what happens" for one person versus another can be seen in and through the charts. Gender, race, Indigeneity, class, sexuality, disability status, and gender identity and expression all overlay what we find in the charts and how these texts are made to matter, even as specific identities may be inconsistently or incorrectly assigned (Daley et al., 2012). While the identities and perceptions of those who write the charts almost always recede into the background of what is written, the intersectional identities and attributed descriptions of those who are documented frame each entry.

Social sorting categories such as gender, race, sexuality, ability, and class are a part of "what happens" in psychiatric charting: they are part of the interactions between people before the writing, the way that people write, what happens when the chart is read, and what other people do with it afterward. While the findings described in this chapter draw upon the larger body of charts and interviews, the examples emphasize particular elements of the data, especially as these relate to the forms of the charting practices themselves (see Chapter 1, Methods). In addition to the 161 charts reviewed in this project, this chapter also draws on data from interviews with 11 participants: eight were conducted as part of the pilot to this larger project, and three were conducted during the main phase of the project (see Chapter 1, Methods). Participants were selected to represent a range of different departments across different hospital units, including those involved in frontline work and others responsible for policy development and training. Interview participants were usually, but not always, also regulated healthcare professionals (e.g., nurses, social workers). These interviews were semi-structured, guided by questions about what the individual saw as key issues influencing charting practices, the roles of organizational and regulatory policies, and protocols regarding safety (e.g., responses to accusations, fears, and experiences of violence). This chapter is not a summary of the interview findings, but rather it integrates interview and chart data to illuminate how the practices of charting are organized in the contemporary psychiatric context (McCoy, 2006; Smith, 2005).

The findings from this project suggest that the formats, technologies, and discourses of contemporary mental health practice reproduce gaps and inequities for people within the mental health system, including those doing the writing and reading and, most significantly, those being written about. These inequities far exceed the intentions of individual and organizational practices and show, instead, the effects of contemporary ideologies and structures. In the next section, the electronic template and formats of the charts are examined as a way in which the activities of differently situated MHPs are coordinated at a structural level, with consequences for how different patients are understood and responded to.

Electronic Charting

Charts used to be exclusively physical objects, written by hand, that were kept for a limited period of time on a unit and then sent to be archived in a separate institutional facility. This form of charting goes back to the origins of psychiatry in the late nineteenth century, and researchers can still access these handwritten charts through various state-run and institution-specific archives (see, e.g., Dyck & Deighton, 2017; Wright et al., 2013). The process of retrieving old charts is onerous and time-consuming; it requires filling out request forms to be approved by the required authorities and waiting for the charts to be retrieved. Frequently, the handwriting is difficult to decipher from one practitioner to the next. Any elements in an "old" chart that someone wants to write into the "new" chart have to be carefully deciphered and then copied or paraphrased anew.

Over the last decade, charting practices have been dramatically altered. Electronic charting has come in as an institutional norm, and many practitioners have straddled the transition from paper to digital in recent memory (Lewis et al., 2011). While there continue to be strict restrictions on who is able to access charts, the amount of effort required to look up an "old" chart has been drastically reduced. The quality of the writing is no longer a barrier, apart from a few typos or the use of unfamiliar acronyms or shorthand. Patients can also more readily request access to their files.

This shift in how to access charts means that what has been written by others becomes much more influential, and the charts take on a greater prominence in everyday practice. As a psychologist on an inpatient unit explained regarding the electronic system and chart retrieval:

Interviewee: … now it's very different. So I certainly … sort of utilize the charts much more than I ever did. If I had to go down to medical records and get a chart out, I, I'd you know rarely did that …

Researcher: yeah …

I: But now … it's, it's easier to do that, so it's just I, I think like I said I'm just sort of aware that if in the same way that I look at other people's notes, they're looking at my notes too. (KI-001)

The difference in being able to retrieve what other service providers have written means that impressions, suspicions, conclusions, decisions, and attributions also become easier to transmit. The documentary residue from one person's encounter with another more readily reaches across individuals, programs, and moments in time. As indicated by the interview participant above, this increased access to notes may make professionals more "aware" of what and how they write and situates their own work in a more crowded set of writers and readers. People are writing for an expanded audience: "they're looking at my notes too."

What are the effects of this increased accessibility to electronic charts? On the one hand, it may lead to greater accountability across providers. People may be more careful to follow the strictest policies and the dominant practices of their institutional context and professional regulators. They would know when, and if, they stand out from the dominant perceptions and practices of their colleagues. Legal concerns and areas of possible challenge would be written about with particular care, as institutional surveillance becomes not only possible but likely. Practices that might once have been done with little concern of oversight or repercussion, such as using restraints, are now more easily quantified, investigated, and challenged by interested parties including patient rights advocates, family members, and patients themselves.

The presence of these "other charters/other readers" may be felt not only while writing the chart, but also while interacting with the people being written about. If people are "seen" through the charts before they are met in person, the previous writers' assessments can haunt the actual encounter between the people in the current space.[1] Any assessment can more readily be compared to what came before. Is the patient described as hostile? If this was previously noted, the provider enters the encounter with this as a possible dynamic to watch for. If the patient seems friendly and open, the provider wonders whether this is a change from their previous way of being. The writer of the current moment is aware that their primary audience is in the future, and that these notes will be used to justify clinical actions, including diagnoses, restraints, privileges, and options around admission and discharge.

While texts have always had the power to influence future interactions, this power has only been magnified with electronic charts that are readily found, read, compared, and surveilled. Each impression of one author can more readily be set in a context of a pattern, where versions of who a person is build across the different entries. The narrative of the person thickens and condenses. It is difficult to be written about anew.

In this electronic chart, the audience is assumed to be elsewhere in time and space, and the note writing is called upon to be universally intelligible. How do categorical differences between patients thus become scripted in standardized ways? One interview participant, an administrator who was involved with the organization and regulation of forms, talked about the way in which there had been a "proliferation" of different program forms prior to the full integration of electronic forms (KI-002). The institution had urged the formal approval of all program forms through a centralized body, but with many different programs and units, this did not entirely stop the use of unapproved forms. This interviewee said that this use of variable and local forms diminished with a fully electronic system because it "forces the hand" (KI-002).

One way in which the format of the electronic charts "forces the hand" to create greater standardization is with the use of drop-down menus that start each new note. Beyond name, gender, and date of birth, the listing includes the following categories: age of appearance, ethnicity, body build (e.g., obese, athletic, average), posture, eye contact, dress, grooming, distinguishing features, behavior, motor behavior, speech (rhythm and prosody, rate, volume, amount, articulation, comments—including accents), affect (range/intensity, quality, congruence), thought content (with comments), delusions, and thought process (with comments). A visual, auditory, and categorizing impression of an individual is readily constructed and transmitted as a constant companion to the writing and reading process. While often many of the categorized boxes read "average," "unremarkable," or "no problem identified," the writer is nonetheless given the constant task of assessing, interpreting, and sorting people's appearances and actions. The drop-down menu becomes an omnipresent technology of frequent and standardized surveillance, whatever the intentions or preferences of the MHP might be. Forms, electronic forms, and checklists in particular mean that each episode of charting organizes the thoughts and attention of the writer along pre-set lines.

In sum, there is no way to interact with or write about individuals without also sorting them into and through categories that you believe other readers will readily understand and agree with. This is not a domain for uncertainty, open questions, or the easy re-assessment of previous interpretations. Drop-down menus assert that the author should be able to make the call and that this version of an encounter is objectively determined.

A further limitation of the drop-down menu is that it repeats the assessment of "abnormality" without allowing for why things occur. For patients who are known to experience hallucinations, their behavior continues to be seen as if there were no other stimuli in their experience. These patients are often noted to be "guarded" or "combative" although their responses may be explained by the things they are seeing or hearing (e.g., see my discussion of Lara below). A drop-down format, instead, implies the possible quantification of discrete attributes or symptoms, which can be added but not directly connected or explained across categories. Each notation is read and reproduced separately in itemized lists, but the connections between them—and their potential explanation in the life experiences and systemic gaps an individual has faced—are left up to the narration offered in the descriptive passages of the chart and to the discretion and initiative of the writer. Hallucinations are certainly not the only "missing" explanations in the charts that ask writers to decontextualize experiences from factors such as racism and poverty, but their absence is especially noteworthy given the emphasis such "symptoms" receive in psychiatric processes and charts.

In terms of social categories with long histories of hierarchy (such as gender, race, class, and sexual orientation), the drop-down list's unassailable categories draw each writer into conversations with dominant beliefs and stereotypes, again, whether this is what they intend or not. The categories that are used are not usually contested, discussed, or modified in the chart, but instead simply reported. The definitions that future readers associate with each category cannot be assessed or responded to. Indeed, these associations extend far beyond the institutional walls. Thus, there is little room for the writer to try to argue against dominant and harmful associations that other people will also bring to each reading. Stereotypes of race and violence/threat are particularly noticeable and consistent across the charts. The charts of racialized patients show particularly strong language that conveys fear or concern for the safety of the MHPs. For example, Charmaine,[2] a 37-year-old, homeless, heterosexual Black woman, was described as follows: "On the ward, Charmaine presented agitated, hostile and belligerent. She belittled staff, affect is angry, menacing and threatening" (psychiatrist). While many behaviors and individuals could be seen as "threatening," the ways in which Black people in particular are "read" and "written" in these psychiatric charts often indicated wariness and fear on the part of the writer (Pilling et al., 2018). Other racialized individuals were also described through a language of "menace," as in the chart of Hina, a 41-year-old, low-income Japanese-Canadian woman:

> Today in her room she was initially not answering any questions Dr. [name redacted] asked. She continued glaring at him menacingly. Pt then continued indicating she was tired of others playing games with her, and requesting she be treated with respect and her well being be taken into account. (social worker)

The language of "menace" has historical associations of seeing people from certain racial groups and parts of the world as foreign and dangerous to the dominant culture and the nation (Goutor, 2007; Thobani, 2007). But such historical associations are implied rather than explicit. The conflict or fear that enters into the encounter is written about as a trait of the patient. It is not put into the context of that individual's life or experience on the psychiatric unit, nor explained as connected to the provider's own feelings and interpretations. For the next practitioner who interacts with the patient, this charting of "threat" will likely shape subsequent interactions and subsequent charting.

Standardization notwithstanding, electronic charts still contain many inconsistencies. For example, physical attributions are not entirely "objective" matters. In one chart, Stacey, a 38-year-old, heterosexual, Black woman, was described as obese, of athletic build, and of normal build, all within a short stay. In another chart, an MHP noted that Alina, a white, heterosexual, 27-year-old woman "looks younger than chronological age" while another wrote that she "looks chronological age." Misgendering was a common form of inconsistency in trans patients' charts. Pronouns sometimes changed within a single sentence, as in this description of Lara, a 61-year-old, white, trans woman who was admitted because of concerns that she was unsafe living on the street in cold weather: "Historically pt. has delusions that her body will combust into flames, and for this reason he douses his clothes with water" (psychiatrist). In this example, it seems that some MHPs were not concerned about misgendering trans patients, even though there is consistent research evidence that misgendering and other transphobic language used by MHPs causes significant harm to trans patients (e.g., Saw, 2017).

Charting inconsistencies can be read as systemic inadequacies. They can also be seen as signs of agency or fallibility in the writers, who are, after all, not infinitely substitutable in their interactions and impressions. Institutions, after all, are made up of the actions of many people, and people seldom act in exact concordance, and inconsistencies show that people who write in psychiatric charts do not always share interpretations. Providers who also inhabit nondominant social positions, such as racialized or LGBTQ providers, may write in ways that resist dominant understandings and attributions, and differ from their white/cis/heterosexual colleagues. It is difficult to know anything about the identities of the providers or whether they deliberately challenged or changed how patients were written about, since provider identities and reflections are excluded from charts (apart from their professional designations). Further, inconsistencies highlight the unreliability of categorizing identities from the perspective of the provider rather than relying on the stated identities of the individual themselves.

Inconsistencies can also be seen in how the charts did, and did not, address trauma histories. This was a highly variable area of charting, with trauma often identified in one section of the chart and absent from others. Throughout the charts reviewed, trauma histories were documented, but this charting seemed disconnected from further practices. MHPs writing on trauma, especially when recent and frequent, primarily documented its existence and accounted for legal obligations (e.g., noting if police were contacted to file a report of assault). Trauma information was seldom integrated into treatment plans or explanatory frameworks. For example, the trauma histories of people with immediate housing concerns were commonly documented but rarely integrated into explanatory frameworks for symptoms and subsequent treatment and discharge plans. Thus trauma appeared throughout these documents without an ongoing connection being made to the person's worldview and relationships.

While formal institutional policies, and professional regulators more generally, consistently assert the need for "trauma-informed practice," interviewees acknowledged that currently this is a work in progress. An awareness of the formats and functions of institutional charting helps us understand the gaps. Trauma appears as a discrete item on the form, often used as a justification for admission, rather than in combination with items surrounding the treatment plan or elements of the daily assessment of symptoms and traits. Only the few providers who find ways to integrate a trauma lens into their narrative descriptions will build and write about these connections; they are otherwise unsupported by the larger format and practice of documentation. Providers who relate to the kinds of trauma recounted by individual patients—cis women, racialized, queer, and trans providers—may be more likely to take this leap.

In addition to the standardized questions and standardized drop-down responses, electronic charts also add the function of copy-and-paste. The ability to quickly copy sections of text from document to document means that even as knowledge about a patient or their situation develops and changes, earlier elements from their chart can re-appear, verbatim. While nonelectronic charting might recycle language from one entry to inspire the next, the effort of looking up an earlier entry and copying it by hand would not be worth it in many situations. In contrast, electronic charts have a built-in incentive to copy passages of text from earlier descriptions, including those by other authors. One clinical practice supervisor noted that copy-and-paste is particularly common for long-stay patients and can result in notes that report daily activities without moving treatment plans forward, possibly prolonging hospital stays (KI-005). In another example, the discharge note for Cindi, a 38-year-old, transgender, Indigenous woman, contained many inaccuracies that can be partly attributed to a copy-and-paste of much earlier notes. In this particular chart, Cindi's reports of transphobic harassment, including "gang threats," in her neighborhood were interpreted as delusional. Her inability to get food because of this fear was the reason given for why she stopped taking HIV medications which require food, while later this decision was simply noted as medication noncompliance without a known cause. Cindi reported no fear or harassment in the hospital and yet her fears regarding her apartment complex continued to be regarded as paranoid. Meanwhile, the chart sometimes used her correct female name and pronouns, and elsewhere misgendered her as "a man," demonstrating that copying and inconsistency can co-exist in electronic charts.

But beyond the question of whether such categories can be consistently and reliably determined by different people at different times, there is a larger principle that cannot be questioned or laid aside: who people are believed to be affects how they are written about. The format of the chart and the requirements of charting practices combine to put intersecting categories of gender, race, sexuality, and class in the foreground. These categories are present throughout each encounter between provider and patient, each moment of providers documenting the encounter, and each experience of reading about a patient through the chart.

Appropriate Concerns: Writing and Imagination

If we lived in a context in which these categories were neutral descriptors, their dominant position in documentation practices and electronic charts would have minimal impact on what happens within and across institutional encounters. However, our society and our institutions have long histories of forming and reinforcing hierarchies across gender, race, class, disability status, and sexuality. As many feminist, critical race, and Mad Studies scholars have pointed out, the history of psychiatry has been intertwined with the dominant Euro-American project of defining "the human" as a white male, often through the rubric of reason.

Psychiatry, from its inception, has been positioned as the nexus through which reason and rationality are to be assessed and asserted, and those designated to be without reason, or "mad," are cordoned off. Communication between those seen as mentally ill and those seen as mentally well is funneled within an institutional, medical framing that positions MHPs as the interlocutors. As Michel Foucault (1988/1961) writes:

> In the serene world of mental illness, modern man no longer communicates with the madman: on one hand, the man of reason delegates the physician to madness, thereby authorizing a relation only through the abstract universality of disease; on the other, the man of madness communicates with society only by the intermediary of an equally abstract reason which is order, physical and moral constraint, the anonymous pressure of the group, the requirements of conformity. (p. 6)

The interaction between the MHP and the patient is not one of free exchange, but always bound by professional requirements to assess, admit, direct treatment, restrain, discharge, and otherwise determine possible increases or decreases in institutional involvement, depending on how far from "reason" the individual seems to be. In addition, this arrangement means that the physician or other MHP must consistently differentiate themselves as a person who possesses reason and can assess it in others. If the provider loosens their own grip on reason, their role as interlocutor dissolves.

Foucault's use of "man of reason" is not accidental, since only white, cisgender men have been seen as legitimate sites of reason in the dominant Euro-American culture. Race is also a founding element of how reason is assigned. Sylvia Wynter (2003) writes:

> [I]t was to be the peoples of the militarily expropriated New World territories … as well as the enslaved peoples of Black Africa … that were made to reoccupy the matrix slot of Otherness – to be made into the physical reference of the idea of the irrational/subrational Human Other. (p. 265)

The Enlightenment project of reason was always primarily a way of sorting people along the lines of race, gender, and reason (Joseph, 2019).

This process of assessing reason is at the core of charting, but "reason" and "rationality" were not the dominant terms in the charts. More commonly, who and what were understandable to the writer were termed "appropriate." In order to write their descriptions of affect, speech, and thought on the chart, MHPs engaged in constant prodding and assessment of how "appropriate" the patient's beliefs, statements, responses, and actions were, leading to long checklists where behavior, speech, affect, thought content, and dress were deemed appropriate or inappropriate. Only in some cases was the context mentioned, as in the frequent description of attire as "Appropriate for the weather." More generally, "appropriate" had no explicit referent but was implied to be objectively unquestionable. Of course, dress, movement, behavior, and speech are highly contextual, and "appropriate" as a term is situational rather than absolute.

Indeed, the mantle of reason and the language of "appropriateness" were consistently pursued and scrutinized by the providers as well, although in a different way. In order to maintain their professional designation, providers needed to do their own work, including documentation, in ways that were "reasonable" and "appropriate"—and thus distinctive from the interpretations of the patients and other nonprofessionals. This language emerged in interviews as providers discussed peer consultations surrounding how and what to chart, discussions that seldom appeared in any of the resulting documentation:

> We do have like we have conversations in terms of like you know should I document this … um, is it or … like you know are we, like I do actually recall in terms of, are we implicating the client if we document this … but I think you know our, *I think the responses though is we need to document but we need to document like appropriately* [emphasis added]. (KI-007)

Some of the fear expressed in the quotation had to do with possible ramifications with other systems—especially courts—if charts were written in ways that caused further problems in the patients' lives. Given their own accountability to the institution, MHPs not documenting something was a risky or unavailable option, but this did not mean that providers were always comfortable with the requirement. Thus "appropriately" is a linguistic bridge, stretching to cover a duty to the client/patient but also a duty to the larger institutional and professional framework; "appropriately" gestures to the limited options the individual providers choose between when their duties conflict.

What seems like appropriate behavior and speech to one provider may not seem this way to the next, and writers used liberal examples of patients' specific words and actions as a way to explain why they assessed them as they did. The attribution of "rational" status is also evident in how explanations for people's experiences were accepted and charted by the providers or disputed and dismissed. Patients with higher class status, and particularly white, middle-class patients, were more often given long narrative explanations for why they appeared the way that they did, especially if there were family members who offered such explanations of behavior (Pilling et al., 2018). For example, Sara, a white, Australian, 35 year-old, middle-class woman was described with a lot of support and understanding:

> Sara stated a lot of judgments about her self regarding her parenting and ability to cope. Writer acknowledged these judgments, and normalized her feelings. Writer also talked about Sara's strength and resiliency that she has shown by getting through these recent months/years. (social worker)

In Sara's chart, different providers documented talking with her about hobbies, interests, and goals of returning to work and traveling, all of which are were given considerable space in the chart alongside discussions of suicidal ideation and behaviors. Sara's husband actively contributed and was described as very supportive, particularly in his willingness to do more of the domestic and childcare tasks while his wife was in the hospital.

Similarly, the chart of Ted, a 57-year-old, white, heterosexual, middle-class man, consistently explained the individual's behavior and symptoms in the context of stressors:

> Wife gives history of a high functioning person interspersed with episodes of mania or depression. … The overwhelming role changes of selling his business and now having to complete and sell his new home. This is fundamentally changing his sense of self. Behavioral activation discussed. (psychiatrist)

This chart does not fall back on terms such as "preoccupation" or otherwise dispute the reasonableness of the patient's explanations of his struggles:

> Ted spoke about his fears and how much they contribute anxiety. According to him, these fears surfaced when he was building his retirement home in [city redacted]. He cited contributing factors as workers not showing up for work and the additional challenges the ones who reported to work faced. Example: having to work harder to complete the job. This client explained that he was unable to cope with these stressors and became sick. (nurse)

Although the provider phrased this explanation in the words of the patient and his wife, little attempt was made to cast doubt on the explanation, particularly as the listing of contextual factors was integrated into an illness framework. Ted was offered electroconvulsive therapy (ECT) but in a way that strongly respected his capacity to make treatment decisions: "made clear patient could cancel referral/withdrawal consent at anytime, as he was having ambivalence regarding procedure" (psychiatrist).

The patient eventually had ECT and was described as "thankful" for the support: "The patient feels safer in hospital and is thankful for the care" (psychiatrist). Gratitude as an attitudinal descriptor was also most likely to appear in charts of middle-class and white patients as was a particular attentiveness to their satisfaction (Pilling et al., 2018).

In contrast, Raquel, a 53-year-old, heterosexual, Black woman living in poverty was not afforded the same explanatory room, especially when she rejected the notion of illness and highlighted systemic inequities:

> According to Raquel she does not have mental illness, but she has "stressors." Specifically, she is a woman, she is poor, black, and does not have enough supports, which causes her to behave the way she does. She is convinced that instead of medications she needs money and education and all her problems would be solved immediately. (psychiatrist)

In this chart, the psychiatrist does not simply report the patient's words but uses phrases to indicate their doubt and disagreement such as

"According to …" and "is convinced." The psychiatrist then followed up by offering a different understanding of the problem:

> I explained her diagnosis, course of illness, prognosis and rationale of treatment in layman's terms, but Raquel dismissed me with a statement, "I will not get better because of medications. I am doing better here because I am surrounded by people." She further assured me she needs a lot of energy to make decisions and to continue on her path of higher education. (psychiatrist)

As Dorothy Smith has described (1978), use of attributed speech in quotation marks for a patient's perspective and neutral speech without quotation marks for the psychiatric writer renders the patient's voice suspicious and dubious. Meanwhile, the specific observations, attributions, and opinions of the writer recede into neutral and objective prose.

People who are seen as experiencing atypical cognition or who have a learning disability can face especially harsh dismissal of their choices and their explanations. Liz, a 43-year-old woman (no race listed, sexual orientation listed as "do not know") was identified as having "mental retardation" and as living in a group home until an outburst of kicking doors led the staff to call police. The chart documents providers' insistent focus on getting Liz to agree to ECT in spite of her repeated objections and concerns regarding memory loss:

> When ECT discussed she expressed unhappiness with the treatments, stating she felt she was forgetting things because of ECT. Client has no insight regarding benefits of ECT in maintaining behaviour to live in the community. Client resorts to regressive behaviour, wanting instead to "live with my Mom." (nurse)

Liz was "encouraged" and eventually agreed to ECT. The follow-up note made it clear that the ECT was seen as a strategy for managing the patient's behavior and was deemed worthwhile even as her fears about memory loss proved true:

> Seen after ECT this am. Pt [patient] is polite, pleasant and essentially vacant. Cannot recall the plans, cannot recall or understand that brother will be picking her up from the boarding home. Cannot remember that she is going to mother's for Christmas. But she is not agitated, hostile or grumpy. When she was reminded she is going to mother's for Christmas her affect and facial expression brightened and she was clearly happy. Pt does have post-ECT confusion. (psychiatrist)

In Liz's chart, we see that the goal of keeping the patient in her community housing even though she would prefer to live elsewhere was used as justification for a treatment that had profound effects on her cognition. The patient's apparently agreeable demeanor was taken as further evidence of the intervention's effectiveness. We are left to wonder if a patient deemed cognitively normal or seen to be reliant on their memory for their work/home life would be urged into an unwanted treatment of this sort.

Seeing patients as "unreliable" or "suspicious" can also transfer across encounters and providers through documentation. The content on "suspicious" patients' charts was often two-fold: the patient was documented as not trusting those around them in the hospital (including other patients and providers), and providers mistrusted what the patient reported. This double-suspicion dynamic was found particularly in charts of patients from marginalized communities who reported concerns about prejudice and violence. For example, Tak, a 63-year-old Asian man reported that other patients were saying racist things about him; this was charted as him being suspicious and experiencing auditory hallucinations. Elsewhere in the chart, providers stated that Tak was potentially untrustworthy in reporting his symptoms and feigning wellness to be discharged. Multiple providers also noted concerns that he was not taking his medications even as he claimed he was and expressed feeling upset at being questioned. These suspicions moved across providers and encounters to coordinate the actions of different providers and the ways that they documented their own interpretation of whether he was trustworthy or not. Further, the patient was charted as "isolative" for not wanting to spend more time with other patients. Throughout the charting, any discussion of racism among the patients got sidelined.

The language and idea of manipulation were associated with certain types of diagnoses, most notably borderline personality disorder but also psychosis, substance use, and depression. For example, Dan, a 24-year-old, white, heterosexual student was given a long list of possible disorders, including personality disorder, and was described as follows in the discharge note:

> In terms of his risk for suicide, he is at chronically elevated risk. While it is clear that he threatens suicide in a manipulative manner (for the past 4 years), at the same time he is quite impulsive, which raises his overall risk, and this is unlikely to change with prolonged admission. (psychiatrist)

In this chart, the language of suspicion was combined with a concern that Dan was not only an unreliable narrator, but that he also willfully shaped a false narrative with an ulterior motive. Manipulation was more commonly attributed to female patients. In one of the provider interviews, a nurse educator answered a question about how gender bias could enter into the charting process with reference to this characterization:

> We're working really hard at … described behaviour, rather than give it a label, that's often, often judgemental. … It used to be very common before, you know, "client is manipulative" and all this stuff. … Just describe the behaviour objectively. Don't give your judgments. (KI-004)

The use of "manipulation" turns the writer's/reader's approach to the patient into one of investigation, and turns all requests and complaints into potential signifiers of pathology. For example, Freya, a 62-year-old, white woman who had been living on the street, was repeatedly described as unreasonable in her requests and concerns: "She was demanding and manipulative, seeking a greater number of passes and wanting us to invest more time in finding her housing 'today'" (psychiatrist).

In the charts examined, it is clear that not every psychiatric patient was equally opposed by "the man of reason" but that all, to differing degrees, were compared with this idealized figure. Some patients were deemed more "appropriate" than others, but all were scanned for actions and words that were out of place, as determined by whether and how the writer was able to understand what the patient said and did. This was a necessarily relational process of assessment, a comparison with what the provider might understand for themselves or interpret in the context of what they knew of the patient. But the extent to which the patient was "foreign" to the provider and "foreign" to the logical models of psychiatry's colonial framing, the less understandable they were likely to be. For patients who were Othered by the overlaying of gender, race, class, disability status, and sexuality, the available pathways to being deemed "appropriate" were more restrictive and overgrown with stereotypes and suspicions.

Conclusion

Too often MHPs, administrators, and researchers see psychiatric charts as inert and residual. This chapter has instead analyzed texts as active elements, lively and full of potential readings and reproductions. It has called attention to the series of social practices that people do through and around texts (Smith, 2005). An analysis of psychiatric narratives alongside ethnographic interviews illuminates the ways in which texts are integrated parts of what people do. As Dorothy Smith (1999) writes:

> The artifice of the text detaches it from the local historicity of living and activity, or seems to do so. But its making was work done in actual settings by one or more people and as part of a course of action, whether of an individual, a group, an organization of some kind, or of an extended social relation coordinating the work of many. And its reading also is in an actual time and place, and enters again into someone's course of action, a speaking part: it becomes active in that course of action. (pp. 135–136)

For example, actions precede the creation of a text (e.g., an intake form) in that people have discussions, write things, discuss further, re-write and negotiate, and transfer texts from other people and places. Actions are also inherent in the creation and "activation" of texts in the process of people writing in particular moments and contexts, but also in the process of individuals or groups reading and responding to (activating) texts in different settings and with different interests and interpretations (McCoy, 1995; Smith, 1999). What makes texts so influential in contemporary social relations is that their recorded nature can separate writing/creation from reading/activation, allowing one text to influence many people across many points in time and vastly different situations (Smith, 2005). Finally, there is further action after the text, which can include decisions or styles of interaction with others that have been influenced by what was read. Often further texts result as a part of this action (e.g., writing a discharge form after reading a series of chart entries and then talking with the person to be discharged). In most institutional settings, text-action chains never end. That is, there are always texts that precede actions and there are always actions prior to texts.

As I have investigated in this chapter, the connections between charts and practices are sometimes difficult to fill in because of the "artifice of the text." But there are enough traces in texts and in our conversations with the people who work with and through texts that patterns emerge. The format developed for electronic charts is more than a neutral substrate, but differentiates patients through features such as drop-down menus, discrete and unassailable categories, and copy-and-paste functions. The integration of interviews and chart data shows that even as MHPs and managers try to work in and through standardized protocols and templates, they are consistently required to assert their own, necessarily variable and potentially inequitable interpretations of patients' behavior, as seen through the language of "appropriateness," "suspicion," and "manipulation." Gender, race, class, disability status, and sexuality are not factored out as mere variables, but are visible in the descriptions and judgments that MHPs make—are required to make. Patients who are closest to social ideals of reason—white, male, cis, heterosexual, able-bodied, middle-class—are written about and responded to as the least suspicious and most understandable. Their needs and desires are more readily respected and considered, their responses seen as appropriate, and their treatment choices considered.

The charts and interviews show us that charting is a structurally determined process, supported by dominant social ideologies and relations that extend far beyond institutional walls. As interview participants explained, policies are not solely determined by individual organizations, but are coordinated across them: "we share policies all the time. You know, we bring up if there is something that our organization is struggling with, who's worked through this issue, we share documents" (KI-009). Charting practices, in particular, are regulated by professional colleges, and all interview participants discussed the ways in which forms were designed around regulatory requirements as well as larger legal mandates. Beyond the nuances of forms and policy, social beliefs rooted in colonialism, heteropatriarchy, and other structural hierarchies furnished the associations and fears that informed what MHPs did and wrote. Thus, as we read psychiatric charts, we are continually confronted by the ways in which oppressive differentiations in how people are categorized are reproduced at the level of structure rather than being limited to the realm of the interpersonal. They must be addressed as such.

Notes

1. Throughout this chapter, I use a variety of terms to describe the people involved in and through the documentation process. Where possible, I use the language of "people" or "individuals," but there are times when it is necessary to understand their institutional roles. I use "providers" or "MHPs" to describe the people in these roles, or sometimes "writers" when referring to the chart itself. When talking about the people whose experiences were documented in the charts, I sometimes use the term "patient" because this is the language of their institutional location, not because I believe this to be an adequate term to describe their full experiences or their identities.
2. All names given to patients are pseudonyms.

References

Daley, A., Costa, L. & Ross, L. E. (2012). (W)righting women: Constructions of gender, sexuality, and race in the psychiatric chart. *Culture, Health & Sexuality, 14*(8), 955–969. https://doi.org/10.1080/13691058.2012.712718

Dyck, E., & Deighton, A. (2017). *Managing madness: Weyburn Mental Hospital and the transformation of psychiatric care in Canada.* University of Manitoba Press.

Flanagan, M., Miller, R., & Davidson, L. (2009). "Unfortunately, we treat the chart": Sources of stigma in mental health settings. *Psychiatric Quarterly, 80*(1), 55–64. https://doi.org/10.1007/s11126-009-9093-7

Foucault, M. (1988). *Madness and civilization: A history of insanity in the age of reason.* Penguin Random House. (Original work published in 1961)

Gibson, M. F. (2016). "This is real now because it's a piece of paper": Texts, disability and LGBTQ parents. *Disability & Society, 31*(5), 641–658. https://doi.org/10.1080/09687599.2016.1197822

Goutor, D. (2007). Constructing the "great menace": Canadian labour's opposition to Asian immigration, 1880–1914. *Canadian Historical Review, 88*(4), 549–576. https://doi.org/10.3138/chr.88.4.549

Joseph, A. J. (2019). Contemporary forms of legislative imprisonment and colonial violence in forensic mental health. In A. Daley, L. Costa, & P. Beresford (Eds.), *Madness, violence, and power: A critical collection* (pp. 169–183). University of Toronto Press.

Lewis, R., Adler, D. A., Dixon, L. B., Goldman, B., Hackman, A. L., Oslin, D. W., Siris, S. G., & Valenstein, M. (2011). The psychiatric note in the era of electronic communication. *Journal of Nervous and Mental Disease, 199*(4), 212–213. https://doi.org/10.1097/NMD.0b013e318216376f

McCoy, L. (1995). Activating the photographic text. In M. L. Campbell & A. Manicom (Eds.), *Knowledge, experience and ruling relations: Studies in the social organization of knowledge* (pp. 181–192). University of Toronto Press.

McCoy, L. (2006). Keeping the institution in view: Working with interview accounts of everyday experience. In D. Smith (Ed.), *Institutional ethnography as practice* (pp. 109–125). Rowman & Littlefield.

Pilling, M. D., Daley, A., Gibson, M. F., Ross, L. E., & Zaheer, J. (2018). Assessing insight: Determining agency and autonomy. In J. M. Kilty & E. Dej (Eds.), *Containing madness: Gender and "psy" in institutional contexts* (pp. 191–214). Palgrave Macmillan.

Saw, C. (2017). Transgender patient care on the inpatient psychiatric unit. *American Journal of Psychiatry Residents' Journal, 12*(11), 7–8. https://doi.org/10.1176/appi.ajp-rj.2017.121103
Smith, D. E. (1978). "K is mentally ill": The anatomy of a factual account. *Sociology, 12*(1), 23–53. https://doi.org/10.1177/003803857801200103
Smith, D. E. (1999). *Writing the social: Critique, theory, and investigations.* University of Toronto Press.
Smith, D. E. (2005). *Institutional ethnography: A sociology for people.* University of Toronto Press.
Thobani, S. (2007). *Exalted subjects: Studies in the making of race and nation in Canada.* University of Toronto Press.
Wright, D., Jacklin, L., & Themeles, T. (2013). Dying to get out of the asylum: Mortality and madness in four mental hospitals in Victorian Canada, c. 1841–1891. *Bulletin of the History of Medicine, 87*(4), 591–621. https://doi.org/10.1353/bhm.2013.0079
Wynter, S. (2003). Unsettling the coloniality of being. *New Centennial Review, 3*, 257–337. https://doi.org/10.1353/ncr.2004.0015

3

Narrating Genders in Psychiatric Inpatient Chart Documentation

Andrea Daley

Introduction

Pointing to the normative base of psychiatry, historical and contemporary critical analyses have implicated psychiatric discourse in constraining individuals within prescribed ways of being gendered (Chesler, 2005; Diamond, 2018; LeBlanc Haley, 2019; Rimke, 2018; Ussher, 2018; Voronka, 2008). Relatedly, psychiatry's adherence to the female-feminine/male-masculine binary and the normal/pathological binary (Daley et al., 2012; Diamond, 2018) has been interrogated as a "mechanism of regulatory control" (Lester, 2013, p. 71). Psychiatric discourse acts as a "restrictive outer narrative" (Hoffman & Hansen, 2017, p. 290) and technology of moral and gendered personhood (Lester, 2013) to

A. Daley (✉)
School of Social Work, Renison University College, University of Waterloo, Waterloo, ON, Canada
e-mail: andrea.daley@uwaterloo.ca

© The Author(s), under exclusive license to Springer Nature Switzerland AG 2021

A. Daley and M. D. Pilling (eds.), *Interrogating Psychiatric Narratives of Madness*,
https://doi.org/10.1007/978-3-030-83692-4_3

limit the range of gendered responses to distress. Moreover, transgressions of normative gender are pathologized in the context of distress. Much of the feminist theoretical and research literature that has explored the relationship between psychiatry and gender has focused on psychiatry's longstanding project of disciplining and regulating cis women with respect to gender performance and sexuality (Rimke, 2018); as Lester (2013) reminds us, psychiatry is "historically and culturally predisposed to find women defective and sick" (p. 71). In this regard, Ussher (2018) not only points to the psychiatric pathologization of femininity but also its participation in regulating gender relations:

> Women who conform to the feminine role and, paradoxically, also those who reject it, were likely to receive a psychiatric diagnosis. At the same time, definitions of mental health were found to coincide with definitions of [cis male] masculinity, whereas femininity was seen as psychologically unhealthy. (p. 73)

Psychiatry's fixation with normative performances of gender and a normative gender order is also evidenced by its longstanding iterations of gender identity disorder and gender dysphoria. Trans and queer critiques of these classifications point to a dichotomized and uniform notion of gender on which they rely, citing such disorders as ideologically defined (i.e., value-based and moralistic) (see Daley & Mulé, 2014). The deeply rooted female-feminine/male-masculine binary underlying these disorders serves as a means to achieve heterosexuality, or rather, to avoid or prevent homosexuality (Hill et al., 2005), as

> structural and institutionalized adherence to rigid dichotomous categories of genders within patriarchal society perpetuates the valorization of Western [white] masculinity while satisfying a misogynistic attempt to inoculate boys and men from acting like girls and women. (Daley & Mulé, 2014, p. 1304)

Critical interrogations such as these are significant in terms of exposing the historical roots of sexism and misogyny in psychiatry, however, they are limited by heteronormativity and cisnormativity and their primary focus on the experiences of white, middle-class women (e.g.,

Chesler, 2005; Ussher, 2011). In this way, they reproduce femininity as a homogenous construct, while ignoring patriarchy as not only structured by gender but also by race, sexuality, and class. This chapter seeks to address this limitation by exploring the psychiatric regulation of gender more broadly through an intersectional lens that considers multiple genders as they are produced at intersections of gender, sexuality, race, and class.

The intersectional analysis presented in this chapter is informed by social constructionist, critical, and postmodern perspectives to deconstruct a subset of chart documentation research data to reveal and rupture the ways in which hegemonic genders and the gender binary operate in psychiatric narratives of distress. It builds on previous related work that has examined the interpretative nature of psychiatry in relation to (1) psychiatric narratives as a technique of power operating to obscure the social and structural contexts of women's distress (Daley et al., 2012) and (2) relationships between assessments of insight and patients' adherence to biopsychiatric explanations of distress and related "treatments" (Pilling et al., 2018). The analysis is aligned with feminist, critical race, decolonial, and Mad Studies scholarship that underscores the interpretative, political, and regulatory nature of determining and classifying "disordered" appearances, styles, behaviors, and feelings (see, e.g., Burstow et al., 2014; Daley et al., 2019; Diamond, 2018; Kanani, 2011; Kilty & Dej, 2018; LeBlanc Haley, 2019; LeFrançois et al., 2013; Mills, 2017; Rimke, 2018; Tam, 2013; Ussher, 2011, 2018; Van Veen et al., 2018). These interrogations collectively problematize psychiatric discourses for transforming complex individual experiences of distress into symptoms of disorder that ignore or dismiss their connection to dominant social, political, and cultural ideologies, including associated norms and moralities regarding gender and the gender order (Georgaca, 2013; Goicoechea, 2013; Lewis, 2017; Marecek & Gavey, 2013). Hornstein (2013) effectively describes the severing of distress from its social and structural contexts by psychiatric discourse as producing a colonizing discourse that "robs people of their right to understand their minds in other terms" (p. 31). Similarly, Mills (2017) points to psychiatry's coloniality, citing its power to erase "alternative ways of knowing, being, and doing" (p. 114). Importantly then, problematizing psychiatry in these way serves as a strategy of resistance in that it calls into question psychiatric diagnosis and related discursive practices as science-informed medical processes of "discovering a pre-existing entity lying inside the sufferer" (Georgaca, 20132013, p. 57), while drawing our attention to how gender, race, sexuality, and class structure the normative base of psychiatry and the psychiatric institution.

Methods

The key themes presented in this analysis were derived from a full set of data (see Chapter 1, Introduction). However, a subset of charts is used to articulate the themes for this chapter. The process of determining the subset of charts centered income as it cut across gender, race, and sexuality as social identity categories. To begin, I selected all charts that had an income and/or source of income listed ($N = 110$).[1] I categorized these charts based on income greater than $30,000 or less than $15,000.[2] These income categories were chosen because they clearly delineate patients who live above and below the poverty line ($19,930) in the province in which the psychiatric institution is located. Patient charts in each income category were then cross-referenced with other known identity categories: gender (cis women, cis men, trans women and trans feminine, trans men and trans masculine, and gender nonconforming), race (Black, "mixed heritage,"[3] Indigenous, and person of color), and sexuality (lesbian, gay, bisexual, and queer).

All charts that indicated an income of greater than $30,000 were included in this analysis given that they numbered only 20. Of the total number of charts that indicated an income of less than $15,000 ($n = 73$), 15 charts belonging to women, including trans women, and 15 charts belonging to men, including trans men,[4] were purposively selected to ensure representation with respect to gender identity, race, and sexuality. In total, 50 psychiatric charts were included in the subset of charts that support this analysis.

Theoretical Constructs of Genders

Theorists have interrogated dominant constructs of femininity and masculinity, including the examination of hierarchies between women and men, among women, and among men as constitutive of the hegemonic gender order, or binary (Brassel et al., 2020; Bryan, 2020; Carter Andrews et al., 2019; Kelly, 2018; McCann, 2020; Unnever & Chouhy, 2020; Vaughan Curington, 2020). Dominant constructs of gender have been variously theorized as emphasized, hegemonic, normative, and patriarchal femininity (Connell & Messerschmidt, 2005; Hoskin, 2019; McCann, 2020) and hegemonic, patriarchal, traditional, and toxic masculinity (Connell & Messerschmidt, 2005; McCann, 2020).[5] For the purpose of this chapter, I use the contingent terms "hegemonic femininity" and "hegemonic masculinity" to signal opposite and complimentary ways of being gendered. They designate ways that people are "culturally constrained to think and act in particular ways" (McCann, 2020, p. 5), serving as ideal constructs against which the appearances, styles, behaviors, and feelings of all individuals are measured (Hoskin, 2019; McCann, 2020). While some theorists have contested the possibility of hegemonic femininity in a patriarchal society that always positions femininity as subservient to masculinity (Connell, 1987), my adoption of this term is aligned with Schippers (2007), who uses it to indicate that dominant ideals of femininity work together with hegemonic masculinity to reinforce a hegemonic gender binary, or a normative gender order.

Undoubtedly, the parameters of hegemonic femininity and hegemonic masculinity have shifted over time in response to social, political, and economic structures. However, changes in societal expectations, beliefs, and norms about appropriate, respectable, and successful expressions of femininity and masculinity may be a matter of degree rather than substance. Notwithstanding progress with respect to women's independence and autonomy in the social, economic, and political arenas (i.e., labor force contributions, political leadership) they are still held to ideal performances of femininity characterized by heteronormative desire and desirability, passivity, emotionality (i.e., irrationality), and compliance. For example, women continue to be "slut shamed" for behaviors that are judged by society to be sexually provocative or promiscuous (Hoskin, 2019). Similarly, the measures of hegemonic masculinity, most notably physical strength, aggression, fearlessness, self-reliance, and rationality continue to be valorized as indicators of manhood even in the presence of available counter discourses that encourage complex and flexible views of manhood (Brassel, 2020). As such, this chapter is premised on an enduring—long-lasting and fixed—hegemonic gender order that is underpinned by hegemonic femininity and hegemonic masculinity.

Findings

The analysis presented below illustrates how gender is narrated in psychiatric documentation of disorder, "treatment," and recovery, revealing ways in which psychiatry serves to police, protect, and maintain the boundaries of normative constructs of gender and the gender order through two inter-related key themes: (1) *gendered aesthetics* and (2) *recognizable gender = credible narrator*.

A total of 50 charts support this analysis: cis women (21) and cis men (25), trans women (3) and trans men (1); Black (9), Indigenous (3), "mixed heritage" (Black-African and white-North American) (3), persons of color (Costa Rican, Iranian, Israeli, Japanese, Mexican, Vietnamese) (6), and white (29) individuals. Two individuals were identified as lesbian and three were identified as gay. Thirty individuals had incomes less than $15,000, and 20 had incomes greater than $30,000. The length of admission ranged from 4 to 90 days, with individuals receiving a range of diagnoses including major depressive disorder, general anxiety disorder, bipolar affective disorder, schizophrenia, schizoaffective disorder, post-traumatic stress disorder, personality traits, dependent personality disorder, and borderline personality disorder. I adopt an intersectional analysis drawing on the subset of charts to explore gender and gender relations between women and men as well as among women and among men as they are narrated in the psychiatric chart.

Gendered Aesthetics

The body, or bodily aesthetics appear to play a central role in determinations of order and disorder in the psychiatric chart. Abstract descriptors of physical aesthetics such as "well-groomed," "appropriately groomed," "well-dressed" on the one hand, and "disheveled" and "poor hygiene" on the other, were replete across the psychiatric charts as indicators of ordered and disordered minds, respectively. Across all charts, however, documentation of the bodily aesthetics of cis and trans women were more frequent and more detailed, with varying levels of nuance, compared to cis and trans men.

Hegemonic Femininity

Bartky (1997) locates femininity as an achievement facilitated by disciplinary practices that "produce a body which in gesture and appearance is recognizably feminine" (p. 27). However, the achievement of normative gender—or whether one is perceived as recognizably feminine—is determined by the white, cisnormative, heteronormative, and classed gaze of the heteropatriarchal psychiatric institution, among other institutions. This analysis reveals the institutional hyper-surveillance of women's own disciplinary practices through the detailed documentation of their physical appearance:

> Age of Appearance: Looks chronological age
> Ethnicity: Asian-East (i.e., China, Japan, Korea) (AEA)
> Body Build: Average
> Dress: Unremarkable
> Grooming: Well-groomed. (Hina,[6] 41 yrs., Japanese, sexual orientation – prefer not to answer, income <15K) (psychiatrist)

> Well groomed, hair neatly kempt, wearing jewelry and clean clothing. Appears her stated age.
> Very thin body habitus. (Brenda, 69 yrs., white, cis woman, heterosexual, income >30K) (psychiatrist)

> Wearing a pink dressing gown. Hair worn long with a big white stripe on the right side. (Victoria, 68 yrs., white, cis woman, heterosexual, income >30K) (nurse)

> Black medium length hair. Neat and clean. Wearing a long dress floral dress. (Vea, 35 yrs., cis woman, variously documented as mixed heritage, Indian-Caribbean, and Black-Caribbean, heterosexual, income <15K, homeless) (nurse)

In the excerpts above, white supremacist ideology that structures the psychiatric institution is invoked as white hegemonic femininity is activated to measure differently racialized bodies during assessments of distress. The documentation of particular bodily aesthetics—body weight, color and condition of hair, the use of make-up, patterns on clothing, and posture(ing)—serves to signal order and disorder. Gender transgressions at the intersection of race and class that threaten patriarchal ideals of modesty, virtue, and fragility (Carter Andrews et al., 2016) are often documented in ways that infer disorder. For example, the documentation of Michele, a Black, cis woman, as having her "thighs exposed" and "messy hair" activates the racist trope of the innately hypersexual, seductive, predatory Black women (Jim Crow Museum, 2020a):

> This is an African Canadian female, she looks older than her age, wearing hospital gown, sitting in bed with her legs up to her thighs exposed.
>
> 60 year old female who looks her stated age. She appeared slightly disheveled with messy hair. (Michele, 60 yrs., cis woman, Black-Caribbean, heterosexual, income <15K) (psychiatrist)

While it is noted in other areas of the chart that Michele is experiencing "burning" and pain in her legs, the documentation narrative from which the first excerpt is extracted fails to make a connection between this sensation and the exposure of her thighs.

A closer look at documentation syntax further illuminates the activation of white hegemonic femininity in specifically raced and classed ways through the binding together of descriptors of bodily aesthetics and descriptions of mood/affect and behavior:

> Ms. [name redacted] presented as an overweight, neatly groomed woman. She was polite, respectful and appropriate during the interview. (Janice, 37 yrs., cis woman, white, heterosexual, income <15K) (psychiatrist)

> MSE: Obese woman who appears her stated age. Affect is irritable, hostile, menacing and dismissive. She is guarded and suspicious. (Charmaine, 36 yrs., cis woman, Black-Caribbean, heterosexual, income <15K) (psychiatrist)

I note that while both women live below the poverty line, Janice is a white woman living alone in a rented apartment, with the support of parents, and Charmaine is a Black woman, "mother of 3 (all in CAS care)," and "homeless/on the street." The use of "overweight" and "obese" to describe the physical aesthetics of the women raises questions about how sexist and fat phobic attitudes about body size are implicated in psychiatric assessments of women's distress. Moreover, contrasting the use of "overweight" to describe a white woman and "obese" to describe a Black woman raises critical questions about the operation of anti-Black (racist) attitudes about body size in the practice of assessment, more specifically. I also note that while both writers document their observations after their respective interactions with the patients, the use of the passive simple past tense to describe Janice focuses the reader on a discrete action completed in a past social context. Conversely, the use of the active present tense to document Charmaine's mood/affect and behavior does not suggest a discrete, completed action but rather that she is actively and innately "irritable, hostile, menacing and dismissive." This documentation implies deliberate action or willingness on the part of Charmaine, and in doing so, activates the "angry Black woman" trope that characterizes Black women as rude, sassy, ill-mannered, and ill-tempered by nature (Jim Crow Museum, 2020b).

Documentation in both charts treats the body as a "modern measure of tangible progress" (Fannon, 2016, para. 4) and constructs Janice and Charmaine as "neutral, compliant instrument of transcendent will" (Garland-Thomson, 2002, p. 5). The written binding of the physical aesthetics descriptor "overweight" to the mood/affect and behavior aesthetics "polite, respectful and appropriate" activates white hegemonic femininity to suggest a correlation between its performance and the potential/willingness for order. Conversely, documentation that brings Charmaine's physical aesthetics in close proximity to the psychiatrist's assessment of her mood/affect and behavior signals a compromised or unsuccessful performance of white hegemonic femininity and, relatedly, disorder at the intersection of Blackness (race) and abject poverty (class).

A review of the documentation of trans women extends this analysis to reveal institutionalized cisnormativity and cissexism upholding white hegemonic femininity. More specifically, mental health professionals' (MHPs) documentation of gender compliance and transgressions, as illustrated below, suggests that these behaviors should be considered notable to the matter of order and disorder. Some documentation serves to activate white hegemonic femininity by noting compliance with its norms:

> Casually dressed transgendered [sic] female of aboriginal [sic] with tattooed arms, wearing flowery blouse; M–>F aboriginal person, dressed in tight pink clothing. (Cindi, 37 yrs., trans woman, Indigenous, sexual orientation documented as "Other: transgendered M-F," income <15K) (psychiatrist)

Importantly, however, Cindi is only precariously recognizable as feminine within the psychiatric institution as indicated by the use of male pronouns by some MHPs, such as this nurse: "patient taken outside to use *his* accompanied passes." Pronoun inconsistency and contradiction underscores the bumping up of the subjective nature of assessment and cissexist and transphobic MHP attitudes against institutionalized practices of standardization (e.g., mental status examination) that are undergirded by cisnormativity.

Other documentation activates white hegemonic femininity by noting gender transgressions (while also misgendering the patient):

> *He* was wearing a skirt, a wig, and moved in a feminine manner; Pt. is wearing a long blonde wig and a short skirt dressed like a woman but unshaven; Pt. moves in an exaggerated feminine manner. (Irie 25 yrs., trans woman, Black-Caribbean, sexual orientation is documented as "Other: Transgendered M-F," income <15K) (nurse)

As Indigenous and Black people, respectively, Cindi and Irie belong to communities that have long been inferiorized, pathologized, and punished by colonial psy discourse and institutions (Rimke, 2018). These chart excerpts suggest that differently racialized bodies may be differently perceived as recognizably feminine, raising questions about the unevenness of racism, for example, anti-Indigenous and anti-Black racism, structuring the psychiatric institution. Unlike Cindi, who is described as being a (trans) woman (flowery blouse, pink clothing), Irie is described as being "like a woman" (short skirt). Irie's identity as a woman is "undone" through the MHP's use of the simile "like a woman." However, while the documentation excerpts differ in the degree to which Cindi and Irie are recognizably feminine to MHPs, they both illustrate the implicit and explicit ways in which cisnormativity and cissexism uphold white hegemonic femininity in the psychiatric institution. Documentation in both charts infers performances of pathological gender as Cindi and Irie are perceived as performing both compromised femininity ("his," "unshaven") and emphasized femininity ("tight pink clothing," "moves in an exaggerated feminine manner").

Of critical importance to the analysis of the activation of white hegemonic femininity in chart documentation practices is the observation that written accounts of cis and trans women's bodily aesthetics are decontextualized from social and structural forces of violence. Women's bodies are made memorable or remarkable through documentation that fails to account for embodied experiences at the intersection of gender, poverty, and trauma. For example, the excerpts below decontextualize women's bodily aesthetics from the impact of poverty and food insecurity on the body ("pale skin"), financial resources for clothing ("wearing the same sweater"), and institutionalized and practitioner gender normativity, cissexism, and transphobia, and relatedly, the availability of resources to sustain gender transition while institutionalized ("dominant male features"):

> Looks small, about 5' 1". Wearing a black skirt and white tank top with a large black bow sewn onto it. Has long black curly hair. Skin is pale. (Martina, 19 yrs., cis woman, variously documented as mixed heritage, Latin American, and Caucasian, heterosexual, income <15K) (nurse)

> Has short red hair. Wearing the same sweater worn Thurs. night when she was admitted. (Kathy, 65 yrs., cis woman, white, heterosexual, income >30K) (nurse)

> 26 year old transgender from male to female with dominate male features [facial hair is documented in the chart repeatedly]. (Nia, 26 yrs., trans woman, Black-North American, sexual orientation documented as "do not know," income <15K) (nurse)

Hegemonic Masculinity

While men's bodies have been less visible in theoretical interrogations of the psychiatric regulation and discipline of gender, they appear in this empirical analysis of the documentation of bodily aesthetics, although they are less likely than women's bodies to be measured with a high degree of vigilance, perhaps as a function of the patriarchal dictates of psychiatry:

> Appears unshaven and has disheveled hair but is showered. (Dave, 43 yrs., cis man, white, heterosexual, income >30K) (psychiatrist)

> He was tall and of average weight. (Jeff, 38 yrs., cis man, white, gay, income >30K) (psychiatrist)

> Heavyset, 33 year old man, medium height, with blonde hair cut very short. (Jonathan, 33 yrs., cis man, white, heterosexual, income >30K) (nurse)

Despite these details appearing fairly innocuous, documentation of men's physical aesthetics often serve to signal and protect the boundary of white hegemonic masculinity and the gender binary. For example, Jonathan's description as "heavyset" or, rather, having a stocky or stout build, can be read as meeting the physical aesthetics of hegemonic masculinity, which allows for men's physical strength through a "heavyset" build. (e.g., physical strength). This type of descriptor is not seen in written descriptions of women's physical aesthetics. Rather, as previously illustrated, women's bodies are typically assessed in reference to weight based on fat phobic and racist attitudes that undergird white hegemonic femininity (e.g., "overweight" and "obese"). The difference in these descriptors of men's and women's physical aesthetics protects the boundaries of white hegemonic masculinity by signaling allowances for diverse aesthetics of strength among cis men and, at the same time, the forbiddance of female masculinity (i.e., women's physical strength).

The chart analysis suggests that other divergences or transgressions from white hegemonic masculinity such as wearing jewelry, nonnormative hair color and style, and vibrant clothing color are deemed equally relevant to assessments of order and disorder, and therefore, worthy of documentation:

> Piercings on lt. upper ear. (Evan, 23 yrs., cis man, white, heterosexual, income >30K) (nurse)

> Thin build, red haired, blue-eyed, Caucasian male with extensive facial hair, in T-shirt and shorts wearing 10–15 charmed/amulets around neck, 2–5 beaded bracelets around wrists. (Andy, 25 yrs., cis man, white, sexual orientation is not documented, income <15K) (psychiatrist)

> Recent haircut/very short. Longer on top like a brush cut. Has blond streaks. Wearing a long sleeved aqua shirt. (Graham, 26 yrs., cis man, white, heterosexual, income <15K) (nurse)

Each documentation excerpt identifies gender transgressions that infer stereotypes of nonnormative sexuality, thus calling into question the successful performance of white hegemonic masculinity. More specifically, the detailed documentation of bodily aesthetics signal a threat to the normative boundary of white hegemonic masculinity as it is propped up by heteronormativity, namely, queer sexuality.

To summarize, the analysis above reveals that gendered aesthetics are central to psychiatric assessments of order and disorder. One's ability to perform recognizable femininity or masculinity signals the absence of disorder or at least the potential/willingness for order. Conversely, bodily aesthetics that transgress hegemonic genders and performances of compromised genders are interpreted as meaningful evidence of disorder (i.e., the inability to comply with gender norms and normative gender relations). Following this, implicating normative and transgressive gendered aesthetics in determinations of order and disorder protects and maintains the normative boundaries of hegemonic femininity and masculinity and the gender binary. The documentation of specifically raced, sexualized, and classed aesthetics of hegemonic femininity and masculinity reinscribes the racial, sexual orientation, and class power dynamics of the psychiatric institution. Their performances are often differently limited for, or unevenly unavailable to, members of equity-deserving groups, meaning that racialized, queer, and/or poor cis and trans women and trans feminine people are more vulnerable to the regulatory and disciplinary practices of psychiatry.

Recognizable Genders = Credible Narrators

In concert with the theme *gendered aesthetics*, this theme is a powerful indicator of the ways in which hegemonic and marginalized femininities, in particular, are narrated in the psychiatric chart. It speaks to how gender aesthetics and norms are implicated in psychiatric assessments of patients' credibility and allowances for patient agency, and relatedly, access to therapeutic options. More specifically, it articulates the alignment of credibility to recognizable and respectable performances of hegemonic femininity and masculinity, and relatedly, gender normative performances of distress.

Documentation that describes Nia, identified as a trans Black-North American woman as "male" in appearance and having facial hair reifies the gender binary, simultaneously pathologizing and erasing marginalized femininities that live between and beyond it. Through this documentation, the MHP constructs Nia as unsuccessfully performing normative femininity (a marker of disorder), while dismissing the institution's responsibility to ensure that social and medical transition needs are identified/included/implemented in care plans or made available during hospital admissions (it is documented that Nia is taking hormone therapy). Subsequent documentation that describes Nia as reluctant to come out of her room and as "isolated to self" illustrates the severe implications for trans people when psychiatric institutions fail to recognize their detrimental adherence to a normative gender binary. Nia's decision to not socialize with co-patients and staff is pathologized as uncooperative ("refused") and abnormal ("withdrawn"), rather than recognized as an act of self-preservation—preservation of identity and safety—in a context of forced mismatched appearance and gender identity: "He [misgendered] refused to socialize and only comes out for meals. Patient is withdrawn most of the time." There is no indication that the documenting nurse attempted to explore why Nia decided or needed to stay in her room except for the purpose of accessing food. Her behavior, like her appearance, is documented as disordered. In this case, Nia's gender identity is erased through gender normative documentation; she is not given the opportunity to be a credible narrator of her own distress. In other words, Nia's unrecognizable gender made her a non-credible narrator.

Documentation that suggests an implicit association between performances of marginalized femininities beyond that of gender aesthetics and non-credibility are also evident in psychiatric charts, for example, the psychiatric documentation of Lee-Ann, an unemployed 31-year-old, Indigenous, cis, heterosexual woman. The characterization of Lee-Ann and MHPs' responses to her distress must be read through the lens of colonial violence, anti-Indigenous racism, intergeneration violence, and gender-based violence against Indigenous girls, women, and Two-Spirit people (Buller et al., 2019). Throughout the chart, Lee-Ann is characterized as "rude and dismissive," "threatening," and using "foul," and "profane" language. Her expressed desire to not take medication is configured as "defiant" even though her rationale is documented:

> Pt remains hypomanic and very excitable. She remains defiant regarding receiving a long acting antipsychotic. (nurse)

> She does not want to be placed on a depot [injectable medication] and does not want a community treatment order, even if this improves her chance of not being admitted to hospital and improves her mental health and functioning, as she does not like the idea of being controlled. (psychiatrist)

> She noted that she's terrified of needles and doesn't want to get diabetes (which she considers a 1:1 relationship with the injections). (psychiatrist)

Characterizing Lee-Ann as "rude," "dismissive," "threatening," and "defiant" marks her as feminine transgressive—noncompliant with the normative signifiers of white, feminine distress, such as "polite" and "cooperative." Notably, several times at the beginning of Lee-Ann's involuntary admission (she was brought to hospital by police as an "emotionally disturbed person," where she was admitted involuntarily on Form 1), MHPs subject her to four-point restraints and oral chemical restraints as a behavior management strategy.

Lee-Ann is also characterized as less than credible when it came to her claim that her distress is a result of gender violence[7] ("trauma and PTSD"), while providing further rationale for her "dislike" of medication. A social worker documented that:

> Pt. discussed her dislike of current medication (seroquel) as she feels it makes her "bloated and gassy." She indicates all medication has not been helpful to her, and attributes her recent periods of aggressive outbursts and inability to manage her emotions last week to her past trauma and PTSD.
>
> Pt. minimizes away several recent hospitalizations and decompensation in the community, stating it was related to her interpersonal issues[8] with her boyfriend.

A psychiatrist's documentation at the point of admission indicated that Lee-Ann was living with a "physically and emotionally abusive boyfriend." However, Lee-Ann's narration of distress caused by "past trauma and PTSD" is negated by social work documentation that infers her lack of awareness (minimizes "decompensation") and reduces gender violence to "interpersonal issues."

The construction of Lee-Ann as "rude," "dismissive," "threatening," and "defiant" and as a non-credible narrator of distress, and the use of physical and chemical force against her needs to be seen as a significant and specific form of violence because she is an Indigenous woman. Indeed, a social worker noted that "pt. stated she identifies with her Native roots." While stated above, it bears repeating here, the response of MHPs to Lee-Ann must be read through the lens of colonial violence, anti-Indigenous racism, intergeneration violence, and gender violence against Indigenous girls, women, and Two-Spirit people (Buller et al., 2019). It must be read as a product of psychiatry's complicity with colonizing, discursive practices and colonial violence that is infused in all Western institutions (Mills, 2017). It must be read through the settler state's (Canada) longstanding policies and practices that have as their aim to "get rid of the Indian problem" (Rheault, 2011, p. 3), most commonly facilitated through the discrediting and punishing of Indigenous knowledges and ceremonies, the rupture of Indigenous communities and confinement of Indigenous people (i.e., over-representation of Indigenous children, youth, and adults in the child welfare and prison systems), and the use of physical and sexual violence in the destruction of families (i.e., residential schools, 60s Scoop) (Buller et al., 2019; Chapman & Withers, 2019). At the very least, Lee-Ann's fear of needles (she is "terrified") related to "getting diabetes" should have been understood by MHPs in the context of the disproportionate prevalence of diabetes among Indigenous Peoples in Canada because of "the complex interaction of multiple determinants of health, many of which are rooted in colonial processes and structures that have altered socio-economic, political and cultural systems" (Halseth, 2019, p. 5).

From this perspective, I raise the question of whether Lee-Ann could ever be received as a credible narrator in a colonial psychiatric institution. Certainly it raises the broader question of whether women who are less able, willing, or permitted to perform white hegemonic femininity—to be polite and cooperative—are more likely to be punished through psychiatric assessment and intervention. A comparison to Carly, a 28-year-old employed, white, cis, heterosexual woman (income >30K), might suggest that this is the case:

> Patient informed writer that she was concerned that taking IM medication will prevent her from being "creative in the class room." Writer encouraged her to talk to her Doctor before discontinuing medication. She stated that her boyfriend "does not listen to me when I discuss my problems with him, sometimes he laughs." She described her relationship

with her boyfriend as being "a great concern to me because I love him and I do not want to change him, but I want him to listen to me. (nurse)

I note the similarities between Lee-Ann and Carly: age, ciswomen, heterosexual, income >30K, involuntary status during admission, behavior described as "combative, disruptive, explosive, guarded, hypervigilant, impulsive, restless, shouting, threatening—verbally, threatening—physically, uncooperative" (from the drop-down menu on the multidisciplinary assessment form), and documentation of threatening behavior on the unit. I also note their differences: Indigeneity versus white and unemployed versus employed. Documentation suggests that Carly is perceived by MHPs as performing recognizable and respectable femininity ("romantic relationship with a gentleman," loving her boyfriend while not wanting to change him), particularly at the intersection of race and class (whiteness): "In a romantic relationship with a gentleman, but not living together, not married, no children" (psychiatrist). Unlike documentation of Lee-Ann, Carly is documented as a credible narrator of her distress.

In different areas of the chart, Carly is described as having "declined" or "refused" medication rather than as being "defiant": "She was offered PRN medication but declined. She said that she feels like she's being drugged" (nurse). Again, through this documentation, the MHP constructs Carly as a more believable narrator of her medication and distress experiences than Lee-Ann. In response to the concerns expressed by Carly about medication, the nurse "encouraged her to talk to her Doctor before discontinuing medication," suggesting that like her, the doctor would receive her as a credible narrator of distress. While the psychiatrist's response was less confident about Carly's credibility as a narrator, they noted her being "upset" rather than "defiant" and described approaching her with a gentleness not evident in the interactions documented in Lee-Ann's chart:

We again talked about the importance of taking prn medications when she is offered. She was very upset about this, said that she knows when she needs it, although this is completely incorrect based on her history, and I tried to gently explain this to her. (psychiatrist)

My analysis suggests that the therapeutic tone of encouragement and respect evident in documentation by these MHPs is more common in the charts of cis women who successfully perform hegemonic femininity as it intersects with race (white), class (middle-class), and sexuality (straight). Following this, documentation conveys MHPs' respect for their agency in treatment and admission decisions, and relatedly, facilitation of access to a broader menu of therapeutic interventions, including information about community supports. For example, documentation in Silvia's (36 yrs., cis women, white, straight, income >30K) chart suggests that MHPs perceived her to be a normative, white woman with class privilege. There are a number of comments about her appearance, for example, "light makeup, applied lipstick in ER. Hair dyed blonde with 0.5–1 inch roots," "In casual attire. Nails were painted," and "able to smile appropriately when talking about pets" (psychiatrist). She is repeatedly described as "calm and cooperative." She was deemed suitable for a specialized program that included group therapy and psychoeducation, and upon discharge was given a list of community psychotherapy and CBT resources covered by government health insurance as well as peer support options in her community. Similar to that of other white, middle-class, straight, cis women, the analysis of Silvia's documentation reveals an association between expressions and narrations of distress that are aligned with white hegemonic femininity, being received by MHPs as a credible narrator, and having access to institutional and community-based therapeutic intervention options:

> Spoke about various tx [treatment] options. Reviewed SSRI, SNRI, Alternative agents, TCA and MAOI tx. She would like to d/c duloxetine. Increase wellbutrin. Perhaps add fluoxetine. No plans for suicide today. Future oriented. Would like to go to baseball game Sunday. (Silvia, 36 yrs., cis women, white, straight, income >30K) (psychiatrist)

Similarly, documentation in Marilyn's chart, a white, 69-year-old, cis, heterosexual woman (income >30K) described below reveals the psychiatrist's recognition of her agency, engaging her in discussion about treatment options that include "medical and non-medical" options, noting her preference for "1:1 over groups":

> Neatly groomed/dressed, wearing jewelry. Reading +++books/magazines on patio. Pleasant, forthcoming, cooperative. Euthymic in affect and reactive, smiling as appropriate. Organized/articulate/goal-oriented. (psychiatrist)

Time and time again, chart documentation illustrates the ways in which the boundaries of white hegemonic femininity are protected and maintained as women who perform recognizable and respectable femininity are rewarded by being believed, afforded agency, and given access to a range of therapeutic intervention options, while women who perform at the margins of white hegemonic femininity and beyond are subjected to the penalty of being constructed as unbelievable and disavowed agency and access to nonmedical therapeutic interventions. The analysis of psychiatric charts indicates that racialized, Indigenous, low-income, and trans women are more likely to be included in the latter, as evidenced by the paucity of relevant documentation in their charts.

Documentation in men's charts echoes that of women's charts in terms of valorizing recognizable and respectable performances of white hegemonic masculinity (e.g., documentation related to men's concerns about gendered inadequacy: employment-related stress and distress, being the "sole breadwinner," and "not having a girlfriend"). However, documentation also points more generally to greater allowance for men's credibility, agency, and access to a range of therapeutic intervention options. For example, unlike the refusal by MHPs of Lee-Ann's nonmedical narrative of distress ("trauma and PTSD," "interpersonal relationships with her boyfriend"), the documentation of Stephen's nonmedical narrative of distress appears unquestioned by the MHP, with Stephen (43 yrs., cis man, white, straight, income >30K) being received as a credible narrator:

> Patient reports that this episode started a year ago when he had conflicts at work. He had a verbal altercation with a co-worker that has continued intermittently. He states that in general he is "sensitive" to confrontation and this makes him feel weak and inadequate. He reports that this dates back to when he was a child and had a significant confrontation with a classmate who threatened to kill him and started dragging him to a bathroom, however Stephen was rescued by a teacher. (psychiatrist)

Importantly, Stephen, described as "athletic," "well-groomed," and "appropriate, calm, cooperative," reported distress related to living up to hegemonic masculinity: "he began evaluating his life and lamenting the fact that at the age of 43 he is unmarried and has not accomplished all the things he has set out to do" (psychiatrist). In other words, his distress narrative explicitly and appropriately endorsed behavioral goals that were aligned with hegemonic masculinity. Similar to the women noted above who performed recognizable and respectable hegemonic femininity, Stephen's agency was encouraged by the psychiatrist with respect to decision-making about a range of institutional and community-based therapeutic options, even in the context of "refusing antidepressants":

> He has been refusing antidepressants but writer provided counseling around this and suggested trying Mirtazapine which would help him sleep and act as an antidepressant. He agreed to consider it. (psychiatrist)

> Patient is open to the idea of therapy, which he has never tried before. He is interested in [name of program] and would like to work on 1) social isolation, 2) developing coping skills – CBT – and 3) creating a sense of routine and planning for the future. (psychiatrist)

The analysis of men's charts also reveals the operation of a white supremacist ideology in the alignment of recognizable white hegemonic masculinity and credibility. Indeed, racialized men perform gender in the context of racism that structures the psychiatric institution. In the excerpt below, for example, the psychiatrist's descriptors of Deiondre's Black body suggest some degree of exceptionality ("well-built" and "dark complexion") to which the reader should attend: "MSE: Today, the patient presented as a well-built 42 year old male of dark skin complexion" (Deiondre, 42 yrs., Black, cis man, sexual orientation—"do not know," income <15K, homeless) (psychiatrist).

This documentation excerpt can be read in two ways. First, the writer chose to include the descriptors because Deiondre performed recognizable white hegemonic masculinity notwithstanding the fact that the exclusion of nonwhite men is foundational to it (Wesley, 2015).

White hegemonic masculinity is theorized as not applicable to racialized men because structural racism prevents their participation in white masculinity (e.g., economic success) (Brassel et al., 2020). However, social institutions such as education, religion, the media, and psychiatry socialize all boys and men to perform in accordance with its logics (Brassel et al., 2020; Bryan, 2020). This may mean that Black men and other racialized men may need to negotiate the social pressure to conform to the ideology of white masculinity (Brassel et al., 2020) or risk the violent consequences of its transgression.[9] As such, the writer's documentation infers Deiondre's strong potential for credibility and agency at the intersection of race and gender.

Perhaps the descriptors were documented because they infer the performance of masculinity beyond the borders of white hegemonic masculinity and thus call into question Deiondre's credibility as a narrator of distress. The descriptors may signal the operation of colorism (Hall, 2015), informing the writer's use of the descriptors "well-built" and "dark skin complexion." This interpretation draws forth the notion that "dark skin is a masculine sign of potency" (Hall, 2015, p. 27) as well as research that indicates that people see Black men as larger and more threatening than white men of the same size (Wilson et al., 2017). Indeed, Deiondre's credibility was deemed questionable as he was assessed as "confrontational" by a psychiatrist after a single, initial meeting. Deiondre presented voluntarily to the hospital with long-standing anxiety, depression, social isolation, and a history of homelessness:

> Interestingly, with seemingly little or no provocation, he was rather irritable and confrontational, accusing the interviewer of being confrontational himself, and stating that he did not like how the interviewer was trying to establish a power dynamic of some sort. At any point when a conversation about expectations in the hospital, such as having to talk to psychiatrists daily and discussing how he was feeling, he assumed a position that he should not be disrespected and became confrontational with the interviewers. (psychiatrist)

The psychiatrist went on to query a personality disorder diagnosis and "trouble" for the "treating team":

> Of note is that the patient seemed to present with a confrontational personality today, which could be a reflection of a more antisocial personality structure, or perhaps a paranoid personality structure (though not necessarily of a psychotic or delusional nature). This may pose a challenge for hospitalization moving forward, and something that his treating team would be prudent to take note of.

These chart excerpts can be interpreted as Deiondre performing exaggerated white hegemonic masculinity characterized by excessive independence and self-reliance. However, a more critical reading raises the question of whether Deiondre truly had access to white hegemonic masculinity as a Black man and whether the psychiatrist's response constituted the activation of anti-Black racism in response to Deiondre's agency. In other words, in the context of the colonial, white supremacist ideologies that structure the psychiatric institution, it is possible that this Black man's agency was read as "rising above his place" instead of "taking his rightful place." Either way it seems as though Deiondre's performance of masculinity was received by the psychiatrist as a non-white masculinity (Kelly, 2018, p. 7), thus constructed as threatening (trouble to the team) and pathologized (personality disorder).

Conclusion

This chapter interrogates how gender, sexuality, race, and class structure the normative base of psychiatry and the psychiatric institution by asking what about gender is most relevant in chart documentation. The analysis draws attention to the psychiatric chart as an institutional artifact rather than an artifact produced by individual MHPs. As such, the analysis is premised on the belief that what is documented in the psychiatric chart represents what MHPs think or know to be most relevant to psychiatric assessment and treatment as defined by institutional policies and protocols.

The theme *gendered aesthetics* suggests that the materiality of bodies and utterances and gestures are intertwined in assessments of distress, particularly for cis and trans women and racialized cis and trans men. Following this, *recognizable gender = credible narrator* indicates that patients who perform white hegemonic femininity and masculinity are rewarded with credibility, agency, and access to intervention options as they are perceived as symbols "of virtue in its own right, tantamount to capability, determination and self-sufficiency" (Fannon, 2016, para. 4). In this sense, white hegemonic femininity and masculinity are inherent in, and essential to, biopsychiatric notions of order. Alternatively, nonhegemonic femininities and masculinities, performances of gender that transgress normative raced, sexualized, and classed ways of being, are punished by being constructed as less than credible, afforded less agency, and deemed less likely to benefit from intervention options.

The intersectional analysis offered through the inter-related themes *gendered aesthetics* and *recognizable gender = credible narrator* points to the central role of white hegemonic femininity and masculinity in psychiatric assessments of distress in ways that go unnoticed and unquestioned. More specifically, the analysis reveals how the psychiatric institution serves as a social and structural mechanism to protect white hegemonic femininity and masculinity and a normative gender order. Of particular importance, the analysis uncovers how documentation practices reflect and reproduce the colonial and white supremacist ideologies of psychiatry and the psychiatric institution. Gendered, racialized, sexualized, and classed norms that structure the psychiatric institution as a colonial institution are reinscribed through documentation practices. By privileging performances of femininity and masculinity "along racial, class, and sexuality based lines" (Dej, 2018, p. 219), the psychiatric institution deploys colonial and white supremacist ideologies, valorizing whiteness (white norms, white culture, and white people), while ordering and disordering patients in a web of hierarchies between women and men as well as among women and among men, as constitutive of normative gender relations.

Acknowledgements The author wishes to thank Dr. Maria Liegghio for her thoughtful review of this chapter.

Notes

1. Not all charts had complete demographic information available. Charts that did not include information about income or income source were excluded from the analysis.
2. The information on income available in the psychiatric charts included the following categories: <14,999; 15,000–29,999; 30,000–59,000; 60,000–89,000, and three categories that accounted for income over $90,000.
3. "Mixed heritage" is a racial category that is available for MHPs to complete on the institution's multidisciplinary assessment form. Examples of "mixed heritage" are offered in parenthesis such as "(i.e. Black-African and white-North American)."
4. Given institutional reliance on the gender binary, the few patients that identified as gender nonconforming were gendered as either female or male. As such, the analysis presented in this chapter "lives" in the gender binary.
5. For a detailed account of various theoretical approaches to gender, see McCann (2020).
6. All names given to patients are pseudonyms.
7. In the majority of psychiatric charts, women's disclosures of gender-based violence were documented, often in Emergency, yet there is no documentation evidence that they were explored or that women's trauma experiences were commonly taken up by the MHPs as a contributor to mental state and/or behaviors (e.g., fear, hypervigilance). For example, Lee-Ann is documented as biting and spitting at a security guard when put in four-point restraints: "Should future codes/restraint events occur, staff should be wary of possible biting/spitting" (psychiatrist). While it is documented that "constant support and reassurance [was] needed and provided" while she was in four-point restraints, there is no documentation that explored the impact of being forcibly restrained given Lee-Ann's history of gender-based and colonial violence.
8. It is unknown whether or not this MHP was aware of previous documentation of Lee-Ann's history of gender-based violence so did not specifically identify physical and emotional abuse or whether the use of the term "interpersonal relationships" is in keeping with the documentation pattern of language use that minimizes women's experiences and disclosures of gender-based violence.

9. It is important to note that while racialized men are held to standards of white hegemonic masculinity, they rarely, if ever, accrue its benefits and power (i.e., dominance).

References

Bartky, S. (1997). Foucault, femininity and the modernization of patriarchal power. In K. Conboy, N. Medina, & S. Stanbury (Eds.), *Writing on the body: Female embodiment and feminist theory* (pp. 129–154). Columbia University Press.

Brassel, S. T., Settles, I. H., Jellison, W. A., & Dodson, C. (2020). Power and race in Black and white men's perceptions and experiences of manhood. *Translational Issues in Psychological Science, 6*(4), 325–343. https://doi.org/10.1037/tps0000257

Bryan, N. (2020). Shaking the bad boys: Troubling the criminalization of Black boys' childhood play, hegemonic white masculinity and femininity, and the school playground-to-prison pipeline. *Race Ethnicity and Education, 23*(5), 673–692. https://doi.org/10.1080/13613324.2018.1512483

Buller, M., Audette, M., Robinson, Q., & Eyolfson, B. (2019). *Reclaiming power and place: The final report of the National Inquiry into Missing and Murdered Indigenous Women and Girls.* National Inquiry into Missing and Murdered Indigenous Women and Girls. https://www.mmiwg-ffada.ca/wp-content/uploads/2019/06/Final_Report_Vol_1a-1.pdf. Accessed 20 November 2020.

Burstow, B., LeFrançois, B., & Diamond, S. (2014). *Psychiatry disrupted: Theorizing resistance and crafting the revolution.* McGill-Queen's University Press.

Carter Andrews, D. J., Brown, T., Castro, E., & Id-Deen, E. (2019). The impossibility of being "perfect and white": Black girls' racialized and gendered schooling experiences. *American Educational Research Journal, 56*(6), 2531–2572. https://doi.org/10.3102/0002831219849392

Chesler, P. (2005). *Women and madness.* Palgrave Macmillan.

Chapman, C., & Withers, A. J. (2019). Assimilation/genocide. In *A violent history of benevolence: Interlocking oppression in the moral economies of social working* (pp. 273–322). University of Toronto Press.

Connell, R. (1987). *Gender and power: Society, the person and sexual politics.* Allen & Unwin.

Connell, R., & Messerschmidt, J. (2005). Hegemonic masculinity: Rethinking the concept. *Gender & Society, 19*(6), 829–859. https://doi.org/10.1177/0891243205278639

Daley, A., Costa, L., & Beresford, P. (2019). *Madness, violence, and power: A critical collection.* University of Toronto Press.

Daley, A., Costa, L., & Ross, L. (2012). (W)righting women: Constructions of gender, sexuality and disorder through psychiatric documentation practices. *Culture, Health and Sexuality, 14*(8), 955–969. https://doi.org/10.1080/13691058.2012.712718

Daley, A., & Mulé, N. (2014). LGBTQs and the DSM-5: A critical queer response. *Journal of Homosexuality, 61*(9), 1288–1312. https://doi.org/10.1080/00918369.2014.926766

Dej, E. (2018). When a man's home isn't a castle: Hegemonic masculinity among men experiencing homelessness and mental illness. In J. Kilty & E. Dej (Eds.), *Containing madness: Gender and "psy" in institutional contexts* (pp. 201–229). Palgrave Macmillan. https://doi.org/10.1007/978-3-319-89749-3

Diamond, S. (2018). Trapped in change: Using queer theory to examine the progress of psy-theories and interventions with sexuality and gender. In B. M. Z. Cohen (Ed.), *Routledge international handbook of critical mental health* (pp. 89–97). Routledge.

Fannon, T. (2016). Out of sight, still in mind: Visually impaired women's embodied encounters of ideal femininity. *Disability Studies Quarterly, 36*(1). https://doi.org/10.18061/dsq.v36i1.4326

Garland-Thomson, R. (2002). Integrating disability, transforming feminist theory. *Feminist Disability Studies, 14*(3), 1–32.

Georgaca, E. (2013). Social constructionist contributions to critiques of psychiatric diagnosis and classification. *Feminism & Psychology, 23*(1), 56–62. https://doi.org/10.1177/0959353512467967

Goicoechea, J. (2013). Invoking and inscribing mental illness: A discursive analysis of diagnostic terminology in inpatient treatment planning meetings. *Feminism & Psychology, 23*(1), 107–118. https://doi.org/10.1177/0959353512467973

Hall, R. (2015). Dark skin, Black men, and colorism in Missouri: Murder vis-à-vis psychological icons of western masculinity. *Spectrum: A Journal on Black Men, 3*(2), 27–43. https://doi.org/10.2979/spectrum.3.2.27

Halseth, R. (2019). *The prevalence of type 2 diabetes among First Nations and considerations for prevention*. National Collaborating Centre for Aboriginal Health. https://www.nccih.ca/docs/health/RPT-Diabetes-First-Nations-Halseth-EN.pdf. Accessed 6 November 2020.

Hill, D., Rozanski, C., Carfagnini, J., & Willoughby, B. (2005). Gender identity disorders in childhood and adolescence: A critical inquiry. In D. Karasic & J. Drescher (Eds.), *Sexual and gender diagnoses of the Diagnostic and Statistical Manual (DSM): A reevaluation* (pp. 7–34). Haworth Press.

Hoffman, G. A., & Hansen, J. L. (2017). Prozac or prosaic diaries? the gendering of psychiatric disability in depression memoirs. *Philosophy, Psychiatry, & Psychology, 24*(4), 285–298. https://doi.org/10.1353/ppp.2017.0041

Hornstein, G. A. (2013). Whose account matters? a challenge to feminist psychologists. *Feminism & Psychology, 23*(1), 29–40. https://doi.org/10.1177/0959353512467964

Hoskin, R. A. (2019). Femmephobia: The role of anti-femininity and gender policing in LGBTQ+ people's experiences of discrimination. *Sex Roles, 81*, 686–703. https://doi.org/10.1007/s11199-019-01021-3

Jim Crow Museum of Racist Memorabilia. (2020a). *The Jezebel stereotype*. Ferris State University. https://www.ferris.edu/jimcrow/jezebel/. Accessed 20 December 2020.

Jim Crow Museum of Racist Memorabilia. (2020b). *The sapphire caricature*. Ferris State University. https://www.ferris.edu/HTMLS/news/jimcrow/antiblack/sapphire.htm. Accessed 20 December 2020.

Kanani, N. (2011). Race and madness: Locating the experiences of racialized people with psychiatric histories in Canada and the United States. *Critical Disability Discourse, 3*, 1–4. https://cdd.journals.yorku.ca/index.php/cdd/article/view/31564/31232

Kelly, C. R. (2018). Emasculating Trump: Incredulity, homophobia, and the spectacle of white masculinity. *QED: A Journal in GLBTQ Worldmaking, 5*(3), 1–27. https://muse.jhu.edu/article/717237

Kilty J., & Dej, E. (2018). Psy, gender, and containment. In *Containing madness: Gender and "psy" in institutional contexts* (pp. 1–12). Palgrave Macmillan. https://doi.org/10.1007/978-3-319-89749-3

LeBlanc Haley, T. (2019). The (un)writing of risk on my mad pregnant body: A mad feminist political economy analysis of social reproduction and epistemic violence under neo-liberalism. In A. Daley, L. Costa, & P. Beresford (Eds.), *Madness, violence, and power: A critical collection* (pp. 184–195). University of Toronto Press.

LeFrançois, B. A., Menzies, r., & Reaume, G. (Eds.). (2013). *Mad matters: A critical reader in Canadian Mad Studies*. Toronto: Canadian Scholars Press.

Lester, R. J. (2013). Lessons from the borderline: Anthropology, psychiatry, and the risks of being human. *Feminism & Psychology, 23*(1), 70–77. https://doi.org/10.1177/0959353512467969

Lewis, B. (2017). Depression memoirs in the circuits of culture: Sexism, canism, neoliberalism, and narrative identity. *Philosophy, Psychiatry, & Psychology, 24*(4), 303–306. https://doi.org/10.1353/ppp.2017.0043

Marecek, J., & Gavey, N. (2013). DSM-5 and beyond: A critical feminist engagement with psychodiagnosis. *Feminism & Psychology, 23*(1), 3–9. https://doi.org/10.1177/0959353512467962

McCann, H. (2020). Is there anything "toxic" about femininity? The rigid femininities that keep us locked in. *Psychology & Sexuality*, 1–14. https://doi.org/10.1080/19419899.2020.1785534

Mills, C. (2017). Global psychiatrization and psychic colonization: The coloniality of global mental health. In M. Morrow & L. Halinka Malcoe (Eds.), *Critical inquiries for social justice in mental health* (pp. 114–137). University of Toronto Press.

Pilling M. D., Daley A., Gibson M. F., Ross L. E., & Zaheer, J. (2018). Assessing "insight," determining agency and autonomy: Implicating social identities. In J. Kilty & E. Dej (Eds.), *Containing madness: Gender and "psy" in institutional contexts* (pp. 191–213). Palgrave Macmillan. https://doi.org/10.1007/978-3-319-89749-3

Rheault, D. (2011). *Solving the "Indian problem": Assimilation laws, practices and Indian residential schools*. https://www.omfrc.org/wp-content/uploads/2016/06/specialedition8.pdf. Accessed 12 January 2021.

Rimke, H. (2018). Sickening institutions: A feminist sociological analysis and critique of religion, medicine, and psychiatry. In J. Kilty & E. Dej (Eds.), *Containing madness: Gender and "psy" in institutional contexts* (pp. 15–29). Palgrave Macmillan. https://doi.org/10.1007/978-3-319-89749-3

Schippers, M. (2007). Recovering the feminine other: Masculinity, femininity, and gender hegemony. *Theory and Society, 36*(1), 85–102. https://doi.org/10.1007/s11186-007-9022-4

Tam, L. (2013). Whither indigenizing the mad movement? Theorizing the social relations of race and madness through conviviality. In B. A. LeFrançois, R. Menzies, & G. Reaume (Eds.), *Mad matters: A critical reader in Canadian Mad Studies* (pp. 281–297). Canadian Scholars' Press.

Unnever, J. D., & Chouhy, C. (2020). Race, racism, and the cool pose: Exploring Black and white male masculinity. *Social Problems, 68*, 490–512. https://doi.org/10.1093/socpro/spaa010
Ussher, J. M. (2018). A critical feminist analysis of madness: Pathologising femininity through psychiatric discourse. In B. M. Z. Cohen (Ed.), *Routledge international handbook of critical mental health* (pp. 72–78). Routledge.
Ussher, J. M. (2011). *The madness of women: Myth and experience*. Routledge.
Van Veen, C., Ibrahim, M., & Morrow, M. (2018). Dangerous discourses: Masculinity, coercion, and psychiatry. In J. Kilty & E. Dej (Eds.), *Containing madness: Gender and "psy" in institutional contexts* (pp. 241–265). Palgrave Macmillan. https://doi.org/10.1007/978-3-319-89749-3
Vaughan Curington, C. (2020). Reproducing the privilege of white femininity: An intersectional analysis of home care. *Sociology of Race and Ethnicity, 6*(3), 333–347. https://doi.org/10.1177/2332649219885
Voronka, J. (2008). Making bipolar Britney: Proliferating psychiatric diagnoses through tabloid media. *Radical Psychology, 7*(2). https://www.academia.edu/3846735/Making_Bipolar_Britney_Proliferating_psychiatric_diagnoses_through_tabloid_media
Wesley, L. (2015). The intersection of race and gender: Teaching reformed gender ideologies to Black males in the context of hegemonic masculinity. *Widener University Journal of Black Sexuality and Relationships, 1*(4), 63–84. https://doi.org/10.1353/brs.2015.0010
Wilson, J. P., Hugenberg, K., & Rule, N. O. (2017). Racial bias in judgments of physical size and formidability: From size to threat. *Journal of Personality and Social Psychology, 113*(1), 59–80. https://doi.org/10.1037/pspi0000092

4

"Slighted and Unheard": The Psychiatrization of Bisexuality

Lori E. Ross and Lucy Costa

Introduction

A significant body of critical scholarship exists that focuses on the ways that lesbian, gay, and transgender identities have been taken up in the discipline of psychiatry. This work has drawn attention to the specific mechanisms through which these identities and associated experiences have been pathologized and psychiatrized, both explicitly (in the form of psychiatric diagnoses) and implicitly (in the form of heteronormative and cisnormative assumptions made about those who access psychiatric services). However, this literature has failed to attend to the ways in

L. E. Ross (✉)
Dalla Lana School of Public Health, University of Toronto, Toronto, ON, Canada
e-mail: l.ross@utoronto.ca

L. Costa
The Empowerment Council, Toronto, ON, Canada
e-mail: lucy.costa@camh.ca

© The Author(s), under exclusive license to Springer Nature Switzerland AG 2021

A. Daley and M. D. Pilling (eds.), *Interrogating Psychiatric Narratives of Madness*,
https://doi.org/10.1007/978-3-030-83692-4_4

which bisexual identities and experiences may be uniquely experienced and pathologized in psychiatric settings and discourse. This is a significant gap given that bisexual individuals report the poorest mental health of all sexual orientation groups and thus are particularly likely to access psychiatric services and given that forms of discrimination associated uniquely with bisexuality (i.e., biphobia and monosexism) have been identified.

This chapter addresses this research gap through a content analysis of chart documentation for 12 bisexual individuals who accessed mental health care in a psychiatric hospital. We examined where, when, and how bisexuality was documented in the psychiatric record, as well as where and when it remained invisible. Our analysis yielded three primary themes. In the first, *charting practices and the intelligibility of bisexuality*, we examine how the structure and practice of psychiatric documentation does or does not makes space for bisexuality to be taken up during a psychiatric admission. In the second theme, *social constructions of bisexuality*, we highlight the ways in which discriminatory stereotypes about bisexual people (e.g., as confused or inconsistent in their sexual identities) are activated in encounters with mental health providers (MHPs), resulting in forms of pathologization that are unique to bisexual identity and experience. Drawing from theoretical Mad Studies, we examine the ways in which the psychiatrization of bisexuality and discriminatory beliefs about bisexuality are mutually reinforced, thus illuminating the role of psychiatric discourse in the continued social stigmatization of bisexuality. In the final theme, *un/successful bisexual admissions*, we explore when and how sexual orientation-specific referrals were made and the role of such referrals in patients' perceptions of the success of their admission experience. We close the chapter with a discussion of possible implications of our work for the care provided to bisexual people in psychiatric institutions, education delivered to health care providers, and the practice of chart documentation.

Theoretical Approach

We approach our analysis through a theoretical lens that integrates elements of two disparate fields: Mad Studies and bisexuality studies as taken up in the psy disciplines, including psychology, psychiatry, and the related field of medicine. In this section, we provide a brief overview of each of these disciplines, and then discuss their potential points of integration for understanding the psychiatrization of bisexual people.

Mad Studies as a discipline developed alongside the broader tradition of critical disability studies, together with other disciplines and social movement activism (LeFrançois et al., 2013). A critical disability studies approach views disability, including disability associated with one's mental state, through the lived experiences of individuals rather than through the perspectives of health professionals. Further, critical disability studies draw attention to how societal power relations determine which ways of being are considered normal and which abnormal and thus requiring medical intervention or correction (Reaume, 2014). Mad Studies applies these critiques specifically to the psy disciplines, centering the voices of those who have been labeled mentally ill, to challenge normative, biomedical understandings of mental wellness and distress (LeFrançois et al., 2013). In examining bisexual peoples' experiences of psychiatrization, Mad scholars call us to attend to the ways in which social and political norms dictate particular assumptions about mental distress and how these are subsequently pathologized alongside or in concert with a bisexual identity.

Most studies of bisexuality as it pertains to mental wellbeing and distress have emerged from the fields of psychology and medicine; this research has established that bisexual people report higher rates of a wide range of poor mental health outcomes relative to all other sexual orientation groups (Bostwick & Harrison, 2020; Ross et al., 2017). Further, this body of research has begun to examine the potential mechanisms underlying these mental health disparities, with a particular focus on discrimination as a primary factor (Feinstein et al., 2020; Ross et al., 2017). That is, researchers have noted that bisexual people face not only homophobic and heterosexist discrimination associated with their same-sex attractions and relationships, but also experience biphobia and monosexism, unique forms of discrimination associated with bisexuality (Dyar et al., 2020; Ross et al., 2017). It has been hypothesized that this added load of discrimination, together with the reality that bisexual people face discrimination from both gay/lesbian and heterosexual people, may account for the high rate of mental distress in this population (Lambe et al., 2017; Ross et al., 2017; Salim et al., 2019).

Although this focus on discrimination moves away from a pathologizing, biomedical understanding of the relationship between bisexual identity and mental distress, much of the research conducted in this area, consistent with its psychological or medical orientation, has had as its unit of analysis the individual bisexual person—their experiences of interpersonal discrimination—and how this impacts their experiences of mental distress. Largely lacking from this analysis has been a focused attention on the macro-level forces (the social, political, historical, and other broader contextual factors) that influence, not only the individual bisexual person, but bisexual communities as a whole. In this analysis, we bring together a Mad Studies orientation and individual-level insights that have been drawn from psychology and other psy disciplines in an attempt to address this gap and bring a more fulsome analysis to our examination of the psychiatrization of bisexual people.

To our knowledge, very little work has attempted this integration: writing on bisexuality has largely been absent from Mad Studies (for an exception, see Bostwick & Harrison, 2020), and Mad Studies scholars have taken up issues of sexual orientation in only a limited way (again, for exceptions, see Carr & Spandler, 2019; Daley, 2013; Pilling, 2013, 2019; Spandler & Barker, 2016); Spandler & Carr, 2021). We could identify only a few works in the broader field of critical disability studies that have taken up bisexual identity specifically (e.g., Caldwell, 2010). Thus, in this chapter, we aim to draw novel connections between these two theoretical orientations in order to illuminate both the macro and micro factors that shape bisexual peoples' experiences of psychiatrization. Our analysis attends to the ways in which systems that oppress bisexual people (i.e., heterosexism, monosexism) interlock with systems to oppress Mad people (i.e., sanism) in ways that we theorize will produce unique experiences of psychiatrization for bisexual people.

Methodology

The analysis for this project is based on chart documentation for 12 bisexual-identified patients. These 12 charts were identified to supplement the original dataset for the larger project (see Chapter 1, Introduction) in order to facilitate a bisexual-specific analysis. The charts were identified using the same process as used for the larger project; that is, they were selected sequentially from four inpatient programs until 12 charts for bisexual patients had been selected. However, in the final sample of 12 charts, only two of the four units were represented: a unit characterized by short stays, either followed by immediate discharge or transfer to another unit, and a general psychiatric unit. Of the 12 patients whose charts were included in this analysis, two were women and 10 were men (all presumed cis given the lack of chart documentation to suggest otherwise). One was identified as Indigenous, two as South Asian, and the remaining nine were presumed white. The mean age was 35.5 (range: 18–57 years). The majority of patients were admitted voluntarily, with one white patient brought in by police and one transferred from another hospital under a Form 1 (a 72-hour involuntary admission for psychiatric assessment). Admissions were relatively brief for all 12 patients, ranging from two days to two weeks.

Analysis of the data entailed reviewing all the documentation for each admission in full, first in its entirety for a contextualized understanding and then a second reading during which the analyst flagged any chart content that either explicitly referenced bisexuality (e.g., noted as a demographic characteristic in the Client and Patient Identification section of the chart or flagged as a demographic risk for suicide in the suicide risk assessment) or could be inferred to have a relationship with the patient's bisexuality (e.g., discussion of intimate relationships or experiences of violence or bullying). The flagged content was then analyzed across charts to identify patterns associated with bisexuality in the chart documentation.

Findings

We identified three primary themes related to bisexuality in the chart documentation. In the first, *charting practices and the intelligibility of bisexuality*, we examine how the structure of the psychiatric record, together with institutional charting practices and policies (as observed in the chart documentation), make it (im)possible for bisexuality to be taken up in the context of a psychiatric admission. We analyze how features such as standardized demographic questions and approaches to suicide risk assessment, among others, leave space for bisexuality to be made visible, or alternatively, made invisible, during a bisexual patient's admission.

The second theme, *social constructions of bisexuality*, includes chart content that touches on common stereotypes, beliefs, and other constructions about bisexuality that have been well documented by bisexual scholar/activists (Firestein, 2007; Hutchins & Ka'ahumanu, 2010; Ochs & Rowley, 2009; Rust, 2000). This includes bisexuality as a state of confusion/transition (Israel & Mohr, 2004), the social invisibility of bisexuality (Yoshino, 2000), and social stigma associated with bisexuality, both as it is manifested in external sources (Dodge et al., 2016) and as it manifests itself internally for some bisexual people (Israel et al., 2019). Below we examine how these common social constructions get taken up in particular ways in psychiatric admission.

The final theme, *un/successful bisexual admissions*, characterizes how the taking up (or not) of a patient's bisexuality appears to contribute to the success of a psychiatric admission in terms of MHP-defined and patient-defined indicators of success. In particular, we examine how referrals to services and supports related to bisexuality appear to factor in to the perceived success of a psychiatric admission. In the sections that follow, we discuss these three themes in turn, using de-identified excerpts of chart documentation to illustrate our discussion.

Charting Practices and the Intelligibility of Bisexuality

In this analysis, it became clear that practices of charting as they determine the structure and function of the psychiatric record are of critical importance in the intelligibility of bisexuality for MHPs and the psychiatric institution more broadly. While our analysis examines the structure and function of psychiatric records in a particular institution, we do not believe that what we find is specific to this institution. Rather, it reflects broad trends in psychiatric charting, particularly as it relates to the inevitable shift to electronic medical records. Electronic charting is notable in that it readily creates defined fields, together with drop-down menus, that signal the relevance of particular identities and experiences, including bisexuality. Thus, in this section, we highlight the two specific components of the chart where bisexuality is explicitly made possible, and in turn, the charting practices that realize (or not) this possibility.

The first field in the electronic medical record where bisexuality is made possible is in the "client/patient identification" section of the multidisciplinary initial assessment form, which includes a section on sexual orientation. One of the options in the drop-down menu is bisexual; as such, patients have a first opportunity to disclose their bisexuality on admission. By virtue of our selection process, all 12 patient charts included in this analysis exhibited this identification of bisexuality with the exception of one patient, whose sexual orientation was documented under the "other" option in the drop-down menu as "bisexual?" One can assume that many other individuals who self-identify as bisexual either chose not to disclose their sexual orientation at intake or were never asked this question. However, from the very beginning of the clinical interaction, bisexuality is recognized and made possible by this option in the medical record. Whether this information is then attended to during the admission, however, appears to be dependent on the context of the admission as well as the bisexual-related knowledge and attitudes of the MHPs involved, as discussed in the sections to follow.

The second field where bisexuality can be noted (as part of a broad notation of sexual minority identities) is as part of the suicide risk assessment, which for most patient charts included in this analysis appeared several times over the course of each admission. Under "demographic risk factors," there is a drop-down menu option for "lesbian, gay, bisexual, transgender" (other demographic risk factors that can also be noted are options such as "Caucasian," "male," and "single"). For the bisexual patients whose charts were included in this analysis, their bisexuality was inconsistently noted as a demographic risk factor for suicide; that is, typically it will be noted for some occasions that the suicide risk assessment is performed, but not others. There is no indication in the narrative portion of the charts that this inconsistency reflects changes in the relevance of bisexuality to the specific patient's risk for suicide in that moment, nor is it clear that this option was selected specifically for those patients who explicitly connected their bisexuality to their mental well-being or the reason for their admission. Thus, we suggest that whether or not this option was selected is likely dependent on whether the MHP completing the suicide risk assessment is aware of the patient's sexual orientation (inconsistencies in the chart documentation in this regard, as further described below, suggest that MHPs do not consistently read their colleagues' notations) and, if aware, whether they saw this as relevant to note in relation to the patient's suicide risk. Thus, the suicide risk assessment, like the client/patient identification section of the chart, is structured in such a way that bisexuality can be made visible and relevant, but the extent to which this occurs during the admission depends upon the specific MHP doing the documentation. The range of realities in this regard is explored below.

Social Constructions of Bisexuality

As noted above, research led by bisexual scholars and activists has characterized the ways that bisexuality is socially constructed, identifying stereotypes, beliefs, and deeply engrained social ideas about bisexuality that profoundly impact the experiences and wellbeing of bisexual people (Israel & Mohr, 2004; Ochs & Rowley, 2009; Ross et al., 2017). Specifically, bisexuality has been constructed to be a state of confusion or a transition state to the "healthier" states of lesbian/gay and heterosexuality as well as an identity associated with hypersexual or unhealthy or immoral sexual activities (Ross et al., 2010). In turn, bisexuality, as a healthy, stable identity, is made invisible, and bisexual people are understood to be incapable of healthy, long-term, and/or monogamous relationships (which are held up as the ideal) (Israel & Mohr, 2004; Li et al., 2013; Ross et al., 2010). We sought to examine if and how these social constructions of bisexuality would manifest themselves in psychiatric chart documentation.

Most prominent in the charts we reviewed was the conflation of bisexuality with a confused sexual identity. Take, as an example, Michael,[1] a 39-year-old, unemployed, white man who was voluntarily admitted for risk of suicide. In the client/patient identification section of the chart, his sexual orientation was noted as bisexual, and in intake notes a psychiatrist documented that "he recently revealed to his family he is bisexual, and this is a major stress for him." In this intake chart documentation, implicit links between his bisexuality and his suicidality are made by the charting psychiatrist, for example, a report from his stepmother of him repeating, "don't you know bisexuals kill themselves?" Perhaps as a result "lesbian, gay, bisexual, transgender" is noted as a "demographic risk factor" in the suicide risk assessment completed at intake. However, despite these connections made on intake, his bisexuality is absent from the chart documentation, apart from a single notation made by the pharmacist: "Current stressors include coming out as bisexual and feeling unsupported." The next mention of his bisexuality appears the day before his discharge, when a nurse noted, "When asked about his comment 'bisexual people get killed' he states that 'I understand how these people feel' though he denies identifying himself as bisexual." The inconsistency between this statement and earlier chart documentation was not explored, and in the discharge note, the discharging psychiatrist stated, "he was future oriented and did not endorse any sexual identity confusion on discharge." Thus, in Michael's admission experience, his bisexuality was ignored by the clinical staff until it became "sexual identity confusion," which was noted as resolved, perhaps as evidence to support the appropriateness of his discharge.

There are, of course, other interpretations of Michael's reported denial of his bisexual identity apart from his initial identification being "confusion" associated with his mental state on admission. Of note, immediately following the statement about "sexual identity confusion," his discharge note indicates "He felt slighted and unheard when our formulation of his issues was expressed to him, and he wished to be discharged ahead of his planned discharge." Given this context, an alternative interpretation of Michael's experience might be that given his experience of feeling "slighted and unheard," taken together with the apparent lack of attention to his experience of feeling unsupported in his bisexuality, he determined that the safest or easiest course of action for him would be to deny his bisexuality to his MHPs when it was finally raised again prior to his discharge. This would be consistent with other research that has documented bisexual peoples' experiences of stigmatizing interactions with MHPs and their decisions not to disclose their bisexuality or seek future mental health care as a result (Eady et al., 2011). It is notable that the specific language of "confusion" was applied by the discharging psychiatrist, rather than, for example, a discussion of this shift in self-identification as being a strategic one. Thus, confusion appears to be the most readily available explanation for this patient's experience in the psychiatric institution.

Also apparent in the charts are social constructions of bisexuality as a transition stage; that is, not a long-term, stable identity. This is apparent in the inconsistencies associated with documentation of bisexuality in charts. Paul, a 57-year-old, homeless, HIV-positive, white man brought in by police, was identified as bisexual in the client/patient identification section, and in several places the chart refers to his ex-wife and son as possible sources of support. However, in several other places in the chart, he is described as a "single gay man." Again, it is unclear whether Paul shifted the label he used to describe himself (for strategic or other reasons) or whether the MHPs documenting his admission understood him to now be gay, perhaps as a result of his self-reported involvement in the "gay community" and his HIV positive status, despite his sexual identification as bisexual and documented prior long-term relationship with a woman. As in Michael's case, this disconnect between how Paul identified himself on intake and how he came to be understood over the course of the admission is never taken up in the chart documentation.

Together these examples point to the most prominent social construction of bisexuality apparent in the charts: that of bisexuality as invisible. In seven of the 12 charts, after the initial documentation in the client/patient identification section that the patient is bisexual, their sexual orientation does not get taken up in any meaningful way for the duration of their admission. Roshan's chart is an example. Roshan, a 45-year-old Sri Lankan man who lived in supportive housing, was involuntarily admitted upon transfer from another institution because of a perceived risk for suicide. His bisexual identity was noted on the client/patient identification section, which is notable given that he was documented as being confused during the intake process and unable to answer several of the questions (e.g., source of income, education level). In one of the several suicide risk assessments that were documented during Roshan's two-day admission, "lesbian, gay, bisexual, transgender" was noted as a demographic risk factor. However, in numerous other suicide risk assessments it was not noted, and there were no other notations pertaining to his bisexuality. Thus, in Roshan's case, there was no exploration of any connection between his bisexuality and his experiences of mental distress; his bisexuality was treated as irrelevant to his psychiatric admission despite being one of the few questions he could answer about himself during his initial intake, and despite it being documented as a risk factor for suicidality, which was the reason for his admission to hospital.

This type of invisibility dominates several of the charts: the 37-year-old, white, bisexual woman who was admitted for distress associated with substance use, the 22-year-old, white, bisexual man whose friend brought him into the emergency department in response to her concerns about him hearing voices, the 55-year-old, Indigenous, bisexual woman who was admitted because of "paranoia" that her food, water, and medications were being poisoned. In each case, the patient's bisexuality was noted in the client/patient identification section and then never mentioned again for the duration of the admission. Did these patients see their bisexuality as not relevant to their admission and did not wish for any discussion or support in relation to it? Did they strategically not draw attention to their bisexuality out of concern that it would complicate their admission, given their prior experiences of discrimination and not being believed? Or are these examples of the psychiatric institution making invisible what it cannot understand or does not wish to see?

Un/Successful Bisexual Admissions

Most of the discharge notes included in these 12 patient charts give an explicit account of the un/successfulness of the admission from the perspective of the discharging psychiatrist, and most of them were described as successful: patients were considered to be stabilized (often on medication), the symptoms (including suicide risk) that led to their admission having been sufficiently resolved for the clinical team to feel discharge was warranted. In other cases, the admission appears to have been unsuccessful, for example, when a patient was allowed to leave against the advice of their MHPs. In our analysis, we interrogate whether and how taking up of a patient's bisexuality might contribute to the success (or lack of success) of an admission, and where possible, we attempt to infer the patient's perception of success, in particular in relation to any care received related to their bisexuality.

The most apparent indicator of the institution taking up a patient's bisexuality is documented referrals to sexual orientation-related, and even bisexual-specific, services and supports. We contrast two charts in which such referrals were made, albeit with very different levels of thoughtfulness and intentionality, and examine the possible contributions of these referrals to the admission's success. We then further contrast these experiences with the experience of a patient who requested such a referral but was denied.

The first chart, that of Adam, documents the most fulsome attention to bisexuality of any of the 12 charts. Adam was a single, white man in his late 30s who was employed full-time and voluntarily presented himself to the emergency room with concerns about drug use and hearing voices commanding him to kill himself. In the client/patient identification section, his sexual orientation is noted as "Other: bisexual?" suggesting that from the first interaction, Adam was open about his sexual orientation and that the clinical staff involved made an effort to go beyond asking the required question and documenting the answer (given the completion of the open text field, "bisexual?"). In the "personal/social history" section of the chart, also completed upon admission but by a different MHP (a social worker), it was noted that Adam had "never been in a relationship before but has had a few sexual encounters with males and females when intoxicated in the past." His bisexuality was not noted again until the fourth day of his admission, when he was seen by a psychiatrist who noted, "Spoke much about factors contributing to his episode. Many chronic issues identified including persistent bullying when younger, poor self image, and sexual identity confusion/lack of significant, genuine relationships." The next day, the psychiatrist's progress note indicated that Adam would meet with the social worker that day, among other things, to discuss "LGBT resources." This first meeting with the social worker occurred on day 6, and by our reading, appeared to be a turning point in Adam's admission.

The social worker's note documents discussion of several issues with Adam including his childhood experiences of bullying, stressful job, and goals related to education. However, it also includes this detailed note regarding Adam's sexual orientation:

> The client also discussed his sexual orientation, and being unsure how to prepare to tell his parents about his identity. The client identified to the writer as being attracted to both men and women, but not having a sexual partner in many years. This is something he wants to change, and feels that the stress from his job has prevented him from being in a positive relationship. The writer spoke to the client about LGBTQ groups in the community, and the client was enthusiastic about exploring these.

This is perhaps the most fulsome documentation related to sexual orientation to appear in any of the 12 charts. Although the social worker didn't use the word "bisexual," they appear to have fully and sensitively explored Adam's needs and concerns related to his bisexual sexual attractions.

Indicators of a "successful" admission begin to appear immediately after this first documentation by the social worker. For example, in the psychiatrist's progress note the next day, it was noted: "Very grateful for care, indicated he wished to volunteer and donate to [hospital] when discharged... feels hopeful for the future." Other notations this same day indicated "Client said he slept well for the first time in years" and "Patient is in good spirits" (charge nurse). In the psychiatrist progress note of the following day, the link between Adam's improvement and the attention to his sexual orientation-related concerns was made explicit: "Future oriented, and feeling very grateful for admission, improvement in symptoms, and support in him being about to be more open and honest in his relationships and to himself."

Sexual orientation-related and bisexual-specific support continued to be offered to Adam during his second visit with the social worker, which was documented as follows:

> The client and writer discussed his sexual orientation and its exploration at length. The writer provided the client information on the [local grassroots bisexual group], the [coming out group for adults] that meets weekly at [local LGBTQ-focused community organization], and some LGBTQ mindfulness-based groups that meet at [community health center with specialized LGBTQ programming]. The client was very appreciative of this information, and talked to the writer about how excited he is about feeling better about himself and his sexual identity, and is looking forward to expanding his social network to meeting new people. He is worried about stigma from his family about his sexual identity, and is not going to disclose this to them for the time being. Writer provided ++ support and validation.

Adam was discharged less than a week after this notation, with LGBTQ- and bisexual-specific referrals exclusively listed as "Community mental health supports" for the client to follow up with. Taken together, the chart documentation suggests that Adam's bisexual orientation was dealt with in a sensitive, appropriate, and comprehensive way, particularly by the social worker, who appeared to be unusually knowledgeable about bisexual community resources and supports. Further, the documentation is suggestive that Adam acknowledged this care in relation to his sexual orientation to be an important contributor to the overall value of his admission and that he left the institution feeling more confident in his sexual orientation and much more hopeful about his goals to build meaningful relationships in his life. By all indicators, both patient-identified (i.e., relationship-related) and MHP-identified (i.e., resolution of symptoms), this admission was highly successful.

A contrasting narrative of a "successful" admission can be seen in the chart of Thomas, who was similarly a single, white man in his early 30s. His primary source of income was the Ontario Disability Support Program, and he was admitted for stabilization of the medication used to treat his bipolar disorder. His sexual orientation was noted as bisexual on the client/patient identification section at admission, but then there was no mention of it for the first full week of his admission (at one point in the chart, his sexual orientation is noted as "Do not know"). Then the following documentation was made by a registered practical nurse:

> Client is stressing about his sexuality, stating he does not know if he will find love. Stated he wants a woman to live with so that he can be sure about his sexuality. Has been on websites sending out requests but has not received any response. Denies SI [suicidal ideation].

Bisexuality is not mentioned, and the patient is now described as unsure about his sexual orientation. The only "action/intervention" documented in response to this disclosure is "therapeutic communication," and no referrals or updates to the portion of the chart where patient-defined goals are recorded were made. A related notation (also by a registered practical nurse) follows shortly after in the chart:

> Client c/o [complained of] feeling anxious, talked about the need to get back on track with his life. Stated he is afraid he will not have a family as in wife and kids. Writer encourage[d] him to focus on the issues that brought him back into hospital then go from there.

In this excerpt, Thomas's concerns related to his sexual orientation and relationship issues appear to be dismissed as irrelevant to the MHP-identified goals of the admission.

Thomas's bisexuality continued to be ignored until a psychiatrist progress note written a few days before his discharge, which, following discussion about his discomfort masturbating on the ward, indicated that Thomas "[a]lso disclosed uncertainty/strife about his sexuality, and thought he may benefit from a coming out group." Although no explicit referral was noted in this documentation, a progress note made on the following day indicated that the patient was to be given a pass to investigate a nearby supportive housing organization and at the same time would stop in at a local LGBTQ center located across the street. Given the lack of explicit attention to Thomas's concerns about his sexual orientation, this referral seems to have been made more as a convenience, given he would be in the neighborhood to explore his housing options anyway. When he returned from his pass, nursing notes indicate that he was "feeling much better, had a good day," and the following day he was reported to be "feeling fantastic."

It is not until the discharge summary that we gain any insight into the potential impact of his referral to the LGBTQ community center on his admission experience. This final psychiatrist note opens as follows: "Patient had a successful admission …." After some discussion of his medication stabilization and impact on mania symptoms, the note moves on to discuss the referrals that were made during admission: "He visited [supported housing organization], but did not like it. He did visit, and plans to attend [LGBTQ community center], as he identified as being bisexual, and wished to explore this further."

In contrast to Adam, clinical staff did not explore Thomas's concerns regarding his sexual identity in any depth. Further, while Adam was deliberately referred to a variety of LGBTQ-focused and bisexual-specific resources, Thomas was referred to the local LGBTQ community center in what appears to be a haphazard way, with no guidance regarding the myriad of programs available and no follow up after his visit. Yet in both cases, Thomas and Adam met the institution's criteria for a successful admission, and by our read, both patients experienced their admissions as successful at least in part because of the LGBTQ- or bisexual-specific resources and supports they were directed to (whether deliberately or in passing).

Further support for our argument that attention to sexual orientation matters in the success of an admission comes from the chart of Fisher, a 25-year-old, white, bisexual man. Fisher began his admission in the context of a voluntary stay at a residential addictions program, where he sought help for his use of crystal meth. His bisexuality was noted only on admission and not taken up elsewhere in the chart; it was also noted that he had a supportive girlfriend. Shortly into his residential stay, Fisher expressed discomfort with his program and asked to be transferred to its LGBTQ section. It was noted that in the group about healthy relationships and sexuality he was currently in, he was quiet and participated very little. Although the request for transfer to the LGBTQ program was documented on multiple occasions, there is no indication in the chart that it was honored or even considered. A few days into his residential stay, Fisher became suicidal and was transferred to a psychiatric unit as an involuntary patient; as soon as the involuntary admission expired, he requested discharge and did not return to the residential addictions treatment program. Given that his total stay in the residential program was only four days long, it is unlikely that this admission was successful on the basis of either patient- or MHP-defined indicators.

Although there may have been logistical reasons why Fisher's request for transfer to the LGBTQ program could not be honored (though these are not documented in the chart), we query whether this heterosexually partnered young man was seen as not LGBTQ enough to warrant admission to the LGBTQ program. Nowhere in the chart are his feelings of discomfort in the program or reasons for requesting transfer to the LGBTQ program explored. However, one can imagine that a bisexual individual might experience discomfort participating in a group on healthy relationships and sexuality with (presumably) predominantly heterosexual people and a request to be transferred would make sense. In this case, patient voice and preference were not centered in the admission, and the outcome was an unsuccessful admission. One wonders how this experience might have unfolded differently had Fisher identified as gay or had been partnered with a man.

As these examples illustrate, admissions to hospitals are often associated with demographic and diagnostic criteria (Lebenbaum et al., 2018) but rarely take into account a wider scope on the nature of the distress and problems that lead to admissions, particularly as they relate to sexual orientation.

Conclusion

In this analysis of 12 medical records documenting the psychiatric admissions of self-identified bisexual people, we have highlighted the ways in which the structure of a psychiatric record, together with the institutional practices of charting and the knowledge and attitudes of individual MHPs, make bisexuality (in)visible in the psychiatric admission. We have examined how common social constructions of bisexuality (e.g., as a state of confusion, as invisible) are reinforced by the institution of psychiatry, as well as how meaningful attention to bisexuality (or in contrast, inattention) contributes to the patient- and MHP-defined success of a psychiatric admission. In this final section, we discuss the limitations of our analysis and close with a discussion of the implications for research, education, and practice, of our findings.

Because of the nature of our data source, we could analyze only the chart records of those patients who self-identified as bisexual at the time of admission, which may contribute to the over-representation of white men in our sample relative to the distribution of bisexual people in the general population. It is possible that those who live at the intersection of other axes of marginalization (because of gender or race, among others) may be less likely to disclose their bisexuality at admission for safety, because of state surveillance, or for other reasons. As a result of this limitation, our analysis cannot address the impact of these intersections in a fulsome way, and more research is needed to examine how various forms of structural oppression acting in concert shape the experience of psychiatric admission for bisexual people. Further, using chart records as the sole data source for this analysis is limiting, given that we could only infer the experiences of patients through the filter of the perceptions of their MHPs and the experiences and actions of MHPs only through what they choose to document in the chart. We appreciate that given not only the medical but also the legal functions of a psychiatric chart there may be much that goes undocumented that could contradict the inferences we have made in our analysis. Additional empirical research using other sources of data, in particular data collected directly from both patients and MHPs, would complement and enhance our analysis.

Despite these limitations, our analysis contributes to an emerging field in which issues of sexual identity and sexuality more broadly are being taken up, through a Mad Studies lens (Daley, 2013; Daley & Ross, 2018; Pilling, 2013, 2019; Spandler & Carr, 2021). Our analysis provides an example of the mechanisms through which psy disciplines reinforce and reproduce damaging social constructions about bisexuality, producing harm for individuals whose experiences do not fit these constructions. In highlighting these mechanisms, our work suggests directions for future theorizing (in further examining issues of sexual orientation and sexuality more broadly through a Mad Studies lens) as well as future empirical work (e.g., exploring the experience of psychiatric admission through the perspective of bisexual individuals themselves, including how they would characterize a successful clinical encounter or outcome). We argue that the relative lack of attention to the experiences of bisexual (and other sexual minority) people is a gap in Mad Studies scholarship that requires further attention, pushing beyond critiques of the pathologization of minority sexual orientations to examine how this pathologization acts as one piece of the larger psy project of upholding sanist values and practices. Mad scholars, therefore, have an opportunity to foster and expand Mad theoretical work to contest the ways in which psychiatric systems perpetuate exclusion of diverse Mad and sexual identities, including bisexuality.

Further, this analysis has important implications for education of MHPs, given that the extent to which bisexuality was taken up during patient admissions appears to be highly variable according to the individual MHP. In turn, our analysis provides evidence that meaningful attention to bisexuality can be important for the patients' perceptions of the success of the admission. In these 12 charts, only one MHP demonstrated knowledge of bisexual-specific resources (making referrals to local bisexual programming), and in multiple cases it was questionable whether providers believed the patient's disclosure of bisexual identity or, at a minimum, considered it relevant to their care. It could be argued that provision of relevant community referrals represents a minimum expectation with respect to providing competent care to sexual minority people; thus, our analysis of these charts suggests there is a great deal more to be done in order to ensure that bisexual people have access to competent and appropriate care during psychiatric admission. There is evidence that MHPs receive little formal education related to sexual orientation (DeVita et al., 2018; White et al., 2015), and within the minimal education that is provided, they may receive none or almost none that is particular to bisexuality. This is so despite evidence that bisexual people have the poorest mental health outcomes of any sexual orientation group (Ross et al., 2010, 2017). Thus, education about these disparities, their mechanisms, and the prevalence of stereotypes and negative beliefs about bisexuality is very much warranted.

Finally, our analysis has implications for the practice of psychiatric charting in that the structure and practice of charting is influential in the extent to which a patient's bisexuality is addressed during an admission. We show the value of including fields in electronic medical records that explicitly make bisexuality visible (e.g., in the client/patient information section or suicide risk assessment) and suggest that other such fields could be of value. For example, the charts of the institution of this study include a section on patient-defined goals; a field to indicate goals related to one's sexual orientation could be of value in maximizing the likelihood that sexual orientation-related concerns are taken up by the clinical team. However, our findings also suggest the need for institutional practices and policies that ensure that the information collected in these fields is taken up in a consistent way by MHPs who are competent to do so. We posit that there is no value—but potential harm—in asking someone about their sexual orientation if this information is not attended to in a meaningful way as it can lead to experiences of homophobia, biphobia, and transphobia. Given the move to electronic medical records and thus widespread access to the information contained within them (including by patients themselves), ethical and appropriate use of sensitive information such as sexual orientation should be prioritized.

Acknowledgements The authors wish to thank Rebecca Bagnarol for administrative assistance in the preparation of this chapter.

Note

1. All names are pseudonyms.

References

Bostwick, W. B., & Harrison, E. A. (2020). Bisexual mental health. In E. D. Rothblum (Ed.), *Oxford handbook of sexual and gender minority mental health* (pp. 269–281). Oxford University Press. https://doi.org/10.1093/oxfordhb/9780190067991.013.25

Caldwell, K. (2010). We exist: Intersectional in/visibility in bisexuality and disability. *Disability Studies Quarterly, 30*(3/4). https://doi.org/10.18061/dsq.v30i3/4.1273

Carr, S., & Spandler, H. (2019). Hidden from history? A brief modern history of the psychiatric "treatment" of lesbian and bisexual women in England. *The Lancet Psychiatry, 6*(4), 289–290. https://doi.org/10.1016/S2215-0366(19)30059-8

Daley, A. (2013). Spaces in place: Negotiating queer in/visibility within psychiatric and mental health service settings. In B. A. Le Francois, R. Menzies, & G. Reaume (Eds.), *Mad matters: A critical reader in Canadian Mad Studies* (pp. 298–308). Canadian Scholars' Press. https://doi.org/10.1111/inm.12038

Daley, A., & Ross, L. E. (2018). Uncovering the heteronormative order of the psychiatric institution: A queer reading of chart documentation and language use. In J. M. Kilty & E. Dej (Eds.), *Containing madness: Gender and "psy" in institutional contexts* (pp. 169–189). Palgrave Macmillan. https://doi.org/10.1007/978-3-319-89749-3

DeVita, T., Bishop, C., & Plankey, M. (2018). Queering medical education: Systematically assessing LGBTQI health competency and implementing reform. *Medical Education Online, 23*(1). https://doi.org/10.1080/10872981.2018.1510703

Dodge, B., Herbenick, D., Friedman, M. R., Schick, V., Fu, T.-C., Bostwick, W., Bartelt, E., Muñoz-Laboy, M., Pletta, D., Reece, M., & Sandfort, T. G. M. (2016). Attitudes toward bisexual men and women among a nationally representative probability sample of adults in the United States. *PLoS ONE, 11*(10), 1–18. https://doi.org/10.1371/journal.pone.0164430

Dyar, C., Feinstein, B. A., Stephens, J., Zimmerman, A. R., Newcomb, M. E., & Whitton, S. W. (2020). Nonmonosexual stress and dimensions of health: Within-group variation by sexual, gender, and racial/ethnic identities. *Psychology of Sexual Orientation and Gender Diversity, 7*(1), 12–25. https://doi.org/10.1037/sgd0000348

Eady, A., Dobinson, C., & Ross, L. E. (2011). Bisexual people's experiences with mental health services: A qualitative investigation. *Community Mental Health Journal, 47*, 378–389. https://doi.org/10.1007/s10597-010-9329-x

Feinstein, B. A., Xavier Hall, C. D., Dyar, C., & Davila, J. (2020). Motivations for sexual identity concealment and their associations with mental health among bisexual, pansexual, queer, and fluid (bi+) individuals. *Journal of Bisexuality, 20*(3), 324–341. https://doi.org/10.1080/15299716.2020.1743402

Firestein, B. A. (Ed.). (2007). *Becoming visible: Counseling bisexuals across the lifespan*. Columbia University Press.

Hutchins, L., & Ka'ahumanu, L. (Eds.). (2010). *Bi any other name: Bisexual people speak out*. Alyson Books.

Israel, T., Choi, A. Y., Goodman, J. A., Matsuno, E., Lin, Y.-J., Kary, K. G., & Merrill, C. R. S. (2019). Reducing internalized binegativity: Development and efficacy of an online intervention. *Psychology of Sexual Orientation and Gender Diversity, 6*(2), 149–159. https://doi.org/10.1037/sgd0000314

Israel, T., & Mohr, J. J. (2004). Attitudes toward bisexual women and men: Current research, future directions. *Journal of Bisexuality, 4*(1–2), 117–134. https://doi.org/10.1300/J159v04n01_09

Lambe, J., Cerezo, A., & O'Shaughnessy, T. (2017). Minority stress, community involvement, and mental health among bisexual women. *Psychology of Sexual Orientation and Gender Diversity, 4*(2), 218–226. https://doi.org/10.1037/sgd000022

Lebenbaum, M., Chiu, M., Vigod, S., & Kurdyak, P. (2018). Prevalence and predictors of involuntary psychiatric hospital admissions in Ontario, Canada: A population-based linked administrative database study. *British Journal of Psychiatry Open, 4*(2), 31–38. https://doi.org/10.1192/bjo.2017.4

LeFrançois, B. A., Menzies, R., & Reaume, G. (Eds.). (2013). *Mad matters: A critical reader in Canadian Mad Studies*. Canadian Scholars' Press. https://doi.org/10.1111/inm.12038

Li, T., Dobinson, C., Scheim, A. I., & Ross, L. E. (2013). Unique issues bisexual people face in intimate relationships: A descriptive exploration of lived experience. *Journal of Gay & Lesbian Mental Health, 17*(1), 21–39. https://doi.org/10.1080/19359705.2012.723607

Ochs, R., & Rowley, S. E. (Eds.). (2009). *Getting bi: Voices of bisexuals around the world*. Bisexual Resource Center.

Pilling, M. D. (2013). Invisible identity in the workplace: Intersectional madness and processes of disclosure at work. *Disability Studies Quarterly, 33*(1). https://doi.org/10.18061/dsq.v33i1.3424

Pilling, M. D. (2019). Changing directions or staying the course? Gender, sexuality, and recovery in Canada's mental health strategy. In A. Daley, L. Costa, & P. Beresford (Eds.), *Madness, violence, and power: A critical collection* (pp. 97–114). University of Toronto Press. https://doi.org/10.3138/9781442629981

Reaume, G. (2014). Understanding critical disability studies. *Canadian Medical Association Journal, 186*(16), 1248–1249. https://doi.org/10.1503/cmaj.141236

Ross, L. E., Dobinson, C., & Eady, A. (2010). Perceived determinants of mental health for bisexual people: A qualitative examination. *American Journal of Public Health, 100*, 496–502. https://doi.org/10.2105/AJPH.2008.156307

Ross, L. E., Salway, T., Tarasoff, L. A., MacKay, J. M., Hawkins, B. W., & Fehr, C. P. (2017). Prevalence of depression and anxiety among bisexual people compared to gay, lesbian and heterosexual individuals: A systematic review and meta-analysis. *Annual Review of Sex Research, 55*(3/4), 435–456. https://doi.org/10.1080/00224499.2017.1387755

Rust, P. C. R. (Ed.). (2000). *Bisexuality in the United States: A social science reader*. Columbia University Press.

Salim, S., Robinson, M., & Flanders, C. E. (2019). Bisexual women's experiences of microaggressions and microaffirmations and their relation to mental health. *Psychology of Sexual Orientation and Gender Diversity, 6*(3), 336–346. https://doi.org/10.1037/sgd0000329

Spandler, H., & Barker, M. (2016, July 1). *Mad and queer studies: Interconnections and tensions* [blog]. Mad Studies Network. https://madstudies2014.wordpress.com/2016/07/01/mad-and-queer-studies-interconnections-and-tensions/. Accessed 26 October 2020.

Spandler, H., & Carr, S. (2021). A history of lesbian politics and the psy professions. *Feminism & Psychology, 31*(1), 119–139. https://doi.org/10.1177/0959353520969297

White, W., Brenman, S., Paradis, E., Goldsmith, E. S., Lunn, M. R., Obedin-Maliver, J., Stewart, L., Tran, E., Wells, M., Chamberlain, L. J., Fetterman, D. M., & Garcia, G. (2015). Lesbian, gay, bisexual, and transgender patient care: Medical students' preparedness and comfort. *Teaching and Learning in Medicine, 27*(3), 254–263. https://doi.org/10.1080/10401334.2015.1044656

Yoshino, K. (2000). The epistemic contract of bisexual erasure. *Stanford Law Review, 52*(2), 353–461. https://doi.org/10.2307/1229482

5

Documenting Restraint: Minimizing Trauma

Juveria Zaheer

Introduction

Chemical and physical restraints and seclusion[1] are coercive practices that are regularly used in psychiatric emergency and inpatient settings (Emmanuel et al., 2013; Jacob et al., 2016). While national data are not consistently available, the Canadian Institute for Health Information (CIHI) reports that in the Canadian Province in which this project was based, over 25% of people (i.e., 1 in 4) admitted to inpatient mental health hospitals experience some form of chemical or physical restraint[2] or seclusion (Emmanuel et al., 2013). Data suggest that dedicated psychiatric facilities (i.e., inpatient mental health hospitals) where staff are trained in verbal and other de-escalation strategies reliably show lower rates of restraint use compared to other health care

J. Zaheer (✉)
Centre for Addiction and Mental Health, Toronto, ON, Canada
e-mail: Juveria.Zaheer@camh.ca

Department of Psychiatry, University of Toronto, Toronto, ON, Canada

© The Author(s), under exclusive license to Springer Nature Switzerland AG 2021

A. Daley and M. D. Pilling (eds.), *Interrogating Psychiatric Narratives of Madness*,
https://doi.org/10.1007/978-3-030-83692-4_5

settings (e.g., general hospitals) (Gaynes et al., 2017; Huizing et al., 2007). Importantly, however, the likelihood of being subjected to a "control intervention" (Jacob et al., 2018, p. 93) is disproportionate across race and class, suggesting the operation of race- and class-based bias in mental health care. For example, evidence suggests that Black male patients are at higher risk of restraint than white male patients, and that homeless individuals are disproportionately subjected to restraint use (Schnitzer et al., 2020). A retrospective chart review conducted in Montreal, Canada has also shown that Black persons of Caribbean or African descent with first episode psychosis were significantly more likely to be coercively treated (including use of seclusion, physical restraints, and intermuscular chemical restraint) (Knight et al., 2021).

Undoubtedly, being subjected to restraint of any kind can be (re)traumatizing and cause significant mistrust of mental health professionals (MHPs) and the mental health care system for individuals (Jacob et al., 2018; Lanthén et al., 2015; Lu et al., 2017; Spinzy et al., 2018; Wong et al., 2020). Jacob et al.'s (2018) exploration of women's experiences of mechanical restraints revealed that "feelings of abuse and violation" were "prominent elements in the recollection of these experiences" (p. 100). The authors emphasized that the violation of the body in the application of mechanical restraints raised questions from women about "ethical conduct, personal rights, and inhumanity in the application of restraints while concurrently exposing their humiliating effects" (p. 101). Not surprisingly, experiences of restraint can negatively affect patients' engagement with mental health care as it is associated with negative feelings and mistrust of staff, resulting in patients being less likely to seek for help and engage with treatment in the future (Khatib et al., 2018; Kontio et al., 2012; Tingleff et al., 2019; Wong et al., 2020).

Research with patients also reveals important process issues that may stem from poor communication between MHPs and patients. For example, many patients cannot identify the reason why they were secluded or restrained as punishment (Kontio et al., 2012), and a major driver of their negative emotion has been identified as the lack of interaction and communication about their restraint, whether before, after, or during it (Kontio et al., 2012; Lamanna et al., 2016; Spinzy et al., 2018).

Some studies of psychiatric inpatients, including patients who have experienced restraint or seclusion, suggest they believe that the use of physical restraints could be justified in certain situations where the risk of violence was clear and imminent and when initiated respectfully and nonpunitively could be justified and even helpful, promoting a sense of calmness and security (Jacob et al., 2018; Kontio et al., 2012; Lanthén et al., 2015; Spinzy et al., 2018). Similarly, some MHPs report that some forms of restraint have some benefit with respect to fostering the safety of patients and staff members and setting behavioral boundaries (Kinner et al., 2017). In one study, MHP respondents described mechanical restraints as "a necessary evil, but a last resort," while acknowledging that the use of restraints can create significant difficulties in developing a trusting therapeutic relationship (Walker & Tulloch, 2020). Many studies indicate that MHPs find the process of restraining patients to be traumatic and distressing (Ling et al., 2015; Walker & Tulloch, 2020).

More generally, large survey datasets and qualitative research indicate that MHPs and service users have similar beliefs about restraint and seclusion, believing that they cause harm, breach human rights, compromise trust, and enact new trauma and trigger previous trauma (Kinner et al., 2017). In one study, the majority of MHPs and service users felt it was both desirable and feasible to eliminate mechanical restraints (Kinner et al., 2017). As such, the inappropriate and overuse of restraints is a major focus of attention for patient advocacy groups, the health care system, and human rights organizations (Allen et al., 2003; Pariseau-Legault et al., 2019; Walker & Tulloch, 2020). This chapter contributes to critical analyses of restraint use by examining the ways in which patients' trauma from being subjected to restraints and MHPs' causing of this trauma is minimized through psychiatric chart documentation practices.

Locating Myself in the Text

As an emergency department psychiatrist, I am ultimately responsible for making the decision whether to enact institutionalized violence against patients through restraint use. This is a decision that I do not take lightly. While everyone with whom I work would prefer that patients were never subjected to the trauma of restraint, we also witness patients' violence directed toward other patients and MHPs. Our commitment to preventing or minimizing harm between patients and between patients and MHPs means that, at times, we make the difficult decision to use restraints to manage a patient's behavior. Critically reflecting on the decision-making processes of MHPs and our biases and complicity in a carceral system, such as a psychiatric institution, transparency and accountability are of utmost importance.

I choose to write in the first person because during my review of the psychiatric documentation data I was struck by the way that MHPs "disappear" into the chart. This happens through documentation patterns such as the use of language to convey patient passivity and asymmetrical reporting. These documentation patterns can be reflective of whether we as MHPs are able to connect with or see ourselves in patients, as well as how we use narrative strategies to absolve our responsibility for institutional violence. Through documentation we remove ourselves from patient narratives of distress. As a psychiatrist, I do not want to see myself or my colleagues causing trauma rather than—or at the same time as—providing urgently needed mental health care. The psychiatric documentation data analyzed in this chapter illustrates this tension.

Methods

Of the 161 charts abstracted for this study, this chapter engages 13 charts that explicitly and implicitly documented use of restraint including chemical restraint (coercive rather than consensual administration of medication to control behavior), physical restraint (the coercive use of devices applied directly or adjacent to a patient's body to reduce physical movement), and seclusion (a type of restraint that involves confining a person in a room). While the 161 charts reveal repeated instances of MHPs engaging strategies to support patients in crisis to avoid restraint and seclusion, I am interested in patient–MHP interactions where traumatic control interventions were enacted. As such, the analysis presented in this chapter does not include charts where documentation indicates only implicit coercive use of restraint, such as the documentation of PRN (as needed) medication that is often presented as a "choice" or is "offered" to patients in distress (e.g., to calm or sedate patients). In this regard, I note that many patients, especially those who are racially and economically marginalized, may not be able to choose not to take or, rather, refuse this medication without traumatic consequences.

The analysis of chart documentation data entailed reading through all 13 chart files that contained de-identified data and chart summaries (see Chapter 1, Introduction). Following this, I identified all documentation data related to restraint use, as well as other relevant documentation of trauma, restraint use debriefing, and the administration of PRN medications (i.e., documentation of implicit chemical restraint). This documentation content was analyzed for key themes related to the documentation of restraint use in psychiatric inpatient units.

Findings

Three key themes were identified through the analysis of documentation data that point to how I will examine the ways in which patients' trauma from being subjected to restraint, and the implication of MHPs' in this trauma, is minimized: (1) *framing patient trauma as disruptive or demanding*, (2) *providers' use of the passive voice and asymmetrical documentation*, and (3) *discounting structural violence*. The minimization of the trauma caused by restraint use is refracted through the lens of gender, race, and class, in this chapter. As the third theme presented in this analysis indicates, in many cases, concerns or "disruptions" from white, middle-class patients were conceptualized as more rational, and MHPs were more likely to use the active voice in documentation.

To contextualize the analysis, I note that in keeping with best practices, the policy of the institution being studied is that restraint without the consent of the patient or their substitute decision-maker is only used in emergency situations and that it should only be used as a last resort when a patient's aggressive or violent behavior presents an immediate risk of serious bodily harm to themselves or others. Restraint should only be used after all reasonable alternatives, less restrictive measures, and de-escalation strategies have been considered or implemented and assessed as not effective. Notably, there is no clear consensus on what is "immediate risk"; different clinicians have different thresholds, and there is no clear definition of "considered or implemented and assessed as not effective." Formally, psychiatrists make the determination on restraint or seclusion use, but practically, the nursing team and psychiatrist make the decision together.

Of the 13 patients whose charts were included in this analysis, five were documented as cis women, seven as cis men, and one as a transgender woman. The mean age of the patients was 42.3 years (20–65 years). Patients were identified as white (5), Black (4), mixed heritage—Indigenous and white (1), mixed heritage—Black and white (1), and Latin American (1) (one chart did not document race). I note the disproportionately high number of Black patients given their relatively low numbers ($n = 22$) in the overall sample of charts (see Chapter 1, Introduction). Eight patients were identified as heterosexual and one as lesbian. The sexual orientation of one patient was listed as "other: transgender." The remaining charts did not list sexual orientation. Nine patients were identified as having an income below $15,000, one above $15,000, and three were listed as "income not known." Reasons for admission included safety concerns (suicide risk, risk of harming others), need for medication or medication stabilization, substance use disorders, symptoms of mood disorders (mania, depression), and psychosis. All patients in this sample were admitted involuntarily under a Form 1[3] of the Mental Health Act and almost all were eventually placed on a Form 3[4] of the Mental Health Act. Length of stay ranged from several days to over two weeks.

Framing Patient Trauma as Disruptive or Demanding

In this section I articulate the ways in which the responses of patients to involuntary hospitalization and restrictive institutional practices and policies (e.g., frequent checks every 15 minutes, forced engagement with MHPs, no smoking policies, and "offering" PRN medication in response to early signs of distress) are documented by MHPs as disruptive or demanding rather than as expressions of trauma, thereby justifying the use of control interventions. Overall, trauma caused by involuntary hospitalization and other confining practices was rarely documented by MHPs as a source of patients' agitation, anger, and fear. For example, documentation of William's[5] (65 yrs., white, cis man, heterosexual, income unknown) behavior implies that he was unpredictably and unreasonably agitated ("without any clear stressor"), thus requiring locked seclusion:

> In the middle of the interview and without any clear stressor, William became precipitously agitated … blocking the door of the l/s [locked seclusion] room, refusing to discuss (treatment) plan any further. (psychiatrist)

However, a critical reading of this documentation illuminates the possibility of two interactional stressors that may have motivated William's response:

> Discussed current (treatment)
> Says that he does not feel he needs any medication … but rather "to be left alone."
> Adherent with Ola [Olanzapine] 10 mg po qhs … however, finds it too sedating.
> We discussed the idea that his current sedation is likely secondary to PRN meds which have been used to contain his aggression.
> Writer proposed a continued titration of olanzapine.
> However, patient declined. (psychiatrist)

First, William was forced to continue a conversation with the psychiatrist after expressing a wish "to be left alone," and second, he was told that the doctor would like to increase the medication dosage against his wishes. Rather than understanding William's distress in the context of these stressors, documentation describes him as unpredictably (read unnecessarily) aggressive, necessitating his locked seclusion and chemical restraint.

Similarly, Devon, a 20-year-old cis man of Indigenous and white descent (heterosexual, income <15K), with a history of intellectual disability, trauma, and substance use was subjected to locked seclusion for verbally and physically (banging on nursing station door, kicking chair) expressing distress in response to being involuntarily hospitalized. There was no documentation that the nurse considered whether his intellectual disability, trauma, and potential experiences of colonial violence impacted his ability to tolerate confinement and communicate distress:

> Patient was also stating he is not crazy, and that he does not belong to this fucking place. He wants to be out with his friend, nothing wrong with his life, but you people pushed medication on him and locked him up making him losing his mind. (nurse)

Other documentation in Devon's chart constructs him as entitled and willfully disruptive in the context of forced confinement:

> After waiting for only a brief time for NSG [nursing] assistance in accessing the internet, Devon began to act out. He became belligerent and physically threatening with staff; kicking the NSG station door and banging on the window, upon returning to his room repeatedly slammed the door. Not able to settle with support. Code white activated and with security present, Devon accepted prn lox 25mg and ativan 2mg po. Still unable to accept any ownership of his b/h [behavior]. (psychiatrist)

Coercive and restrictive hospital practices associated with involuntary admission are also implicated in MHP documentation that characterized patients' trauma responses as disruptive or demanding. Ebo, a 33-year-old, Black, cis man (heterosexual, income <15K) with a recent immigration history was documented as "escalating with agitation over the morning. Asking and demanding for 'my break … let me out to smoke.' NRT [nicotine replacement therapy] offered multiple times, but he refused" (nurse). Ebo was placed on continuous observation, with a staff member following him on the unit. His inevitable frustration with being followed was documented by the provider: "Client spent majority of time pacing around unit. Client became agitated and verbally aggressive toward writer. Client in a hostile tone stated to writer if writer is not going to give him break, writer should stop following him" (nurse). This documentation illustrates how the imposition of repeated unwanted "interventions" by MHPs affect a patient's behavior, which is then responded to with coercive force. Moreover, documentation reveals how restrictive and coercive institutional practices and policies, such as no smoking and continuous observation, are deemphasized in documentation, receding into the background as sources of patients' distress. Instead, MHPs' documentation narratives abstract patients' distress responses from the institutional context, constructing patients as innately demanding, disruptive, or hostile and in need of control intervention.

Of critical importance is that at times responses that do not actually place patients or staff at imminent risk of bodily harm are assessed as disruptive enough to necessitate restraint. This can have detrimental trauma consequences for patients. For example, the documentation below suggests that Monica, a 50-year-old, white, cis woman (lesbian, income <15K), was restrained and given intramuscular (IM) medications in the absence of a clear indication of her being a safety risk:

> Client was noted to be awake all night; pacing+++, disruptive – opening and banging the doors; rearranging the furnitures [sic] at the lounge; putting books in the freezer; moving and turning chairs upside down; going to the male's washroom; she was offered but refused PRNs even with security's assistance; difficult to redirect and she has a lot of demands – single room, to open the TV, activity room; insight and poor judgment noted; MD on call was called – MD from emerg came and assessed client – ordered STAT [immediately] dose of IM PRNs (2mg ativan and 25mg loxapine) – administered with presence of security; client remains disorganized; will continue to monitor and assess. (nurse)

It is important to note that the presence of security is a coercive intervention, and if Monica had taken the medication when she was initially offered it, it would have been documented as "chemical restraint." In other words, if a patient refuses a PRN and security is called for the purpose of enforcing medication adherence, the patient is considered to have been chemically restrained. Many patients, especially those who have been involuntarily held in psychiatric settings, are offered PRN medication in the absence of security. While this is not considered restraint, it may be the first step down a coercive pathway. While the justification for PRN medication is to calm and sedate a distressed patient and to prevent the use of restraint to manage distress responses, a patient's refusal to take the PRN often results in being chemically or physically restrained or secluded.

Undoubtedly, the threat to safety for patients and MHPs in some situations is real. Nursing staff, who are most likely to be women, are most often at risk of physical harm during patient interactions (Itzhaki et al., 2018). However, it is imperative that psychiatric institutions recognize the productive effect of their policies and practices to motivate patients' trauma responses, including agitation, fear, and anger, and subsequently to put patients at risk for control interventions. In this regard, some documentation indicates institutional and MHP recognition of the (re)traumatizing impact of restraints. For example, Juan, a 41-year-old, Latin American, cis man (heterosexual, income <15) with a trauma history was documented as "quite re-traumatized by the restrains" (psychiatrist) and as having experienced seclusion as "more traumatizing and counter therapeutic" (based on restraint use during a previous admission) (psychiatrist). Notwithstanding this recognition and that debriefing following restraint use is part of hospital policy and trauma-informed care, very few charts include detailed documentation on the impact of restraint use on patients. Typical documentation includes "Pt. debriefed following incident," "Debrief refused by client," and "Supportive communication and orientation provided." When trauma debriefs after restraint use are documented, they often focus on having a patient explain their behavior leading up to restraint rather than an exploration of trauma associated with restraint itself. For example, the chart of Sean, a Black cis man (heterosexual, income unknown) includes the following post-seclusion debrief:

1. What do you believe caused the restraint or seclusion?

   I posed a question to Dr. [name redacted] perceived as a treat, I had conviction + was defiant. I did not refuse medication, only desired to speak with doctor first therefore it was not not-compliance. After, the standoff took place, I took my medication in front of all present + submitted to them for seclusion.

2. a) a) What could staff have done differently that might have prevented the restraint or seclusion?
   The fact that I was admitted + kept confined without fresh air for a period of 48 hrs goes against [the institutions]'s mandate.
   b) What could you have done differently that might have prevented the restraint or seclusion?
   Taken the medication as directed.
   c) Could something have been done during the restraint or seclusion that might have helped to end it sooner?
   The compliance of the doctors to my request.
   d) While you were restrained or secluded is there anything staff could have done to help you (eg cover you with a blanket or play music)
   Give me my ipod
   e) Did you sustain a physical injury …?
   No
   f) How has the restraint or seclusion affected you?
   In no way.
3. What was it like for you to be restrained? (this includes being in seclusion …)
   Used to it, had it done during previous admissions to hospital
4. Did you and the treatment team develop a plan of care to help prevent another restraint or seclusion?
   No
5. a) Left blank
   b) Is there anything else the treatment team can do now and over the next few days to help you recover from this event?
   No. (Written by patient and transcribed verbatim)

This practice (documenting a trauma debrief in a structured format) is atypical and not observed in any other chart. I note that in this debrief, Sean asserted his agency in multiple ways including emphasizing that he "did not refuse medication, only desired to speak with doctor first therefore it was not non-compliance" and contextualizing his response in relation to institutional policies that kept him "confined without fresh air for a period of 48 hrs." Of particular concern is the assumption of alliance between the patient and MHP, whereby after experiencing the violence of restraint Sean was expected to work collaboratively with the MHP to explore how being subjected to coercive force impacted him (How has the restraint or seclusion affected you? What was it like for you to be restrained?) and develop a plan of "care" to prevent the use of future force (Did you and the treatment team develop a plan of care to help prevent another restraint or seclusion?). It's not surprising that Sean responded "no" to the question of whether "the treatment team can do anything now and over the next few days to help [him] recover from the event."

Providers' Use of the Passive Voice and Asymmetrical Documentation

The excerpt above from Monica's chart also serves as a segue into this second theme, which illuminates the ways in which MHPs' agency in restraint use and coercive force (i.e., violence) is effaced from the psychiatric chart through documentation. Notably, the nurse documented that Monica refused the IM PRN "even with security's assistance" but then eventually received it "with presence of security." In similar ways, MHPs' use of the passive voice characterizes the documentation of restraint use in other charts, drawing the readers' attention to who is being acted on (the patient) rather than the person who is responsible for the action (the MHP). This is used through commonly used terms such as "present," "placed," "required." In another case, Glenda's (50 yrs., white, cis woman, sexual orientation and income not listed) forced confinement was documented as an intervention that she "required": "Today pt. is labile. Requiring locked seclusion over the weekend" (psychiatrist). Similar documentation excerpts in other charts abound:

> Security were called and she took emergency medication by mouth. She required seclusion for only a short period and slept the majority of the night in her room (Joanna, 28 yrs., white, cis woman, heterosexual, income >30K). (psychiatrist)

> Required restraints and chemical sedation in ER. (Chris, 31 yrs., white, cis man, heterosexual, income <15K) (psychiatrist)

> At 0355 hrs, with three security staffs [sic] present, patient received PRN ativan 2 mg PO and loxapine 25 mg PO. Duty doctor [name redacted], and nursing supervisor present. Locked seclusion ordered. At 0400 hrs, patient placed in unit 2-2 locked seclusion room. (Devon) (nurse)

Use of the passive voice suggests that the MHP is an irrelevant or insignificant actor, thus requiring only a vague reference in restraint use documentation. The effacing of MHPs in chart documentation serves to absolve them—and the institution—of responsibility and accountability for the events leading up to restraint use. The reader is not given details about MHPs' actions, including their knowledge of and ability to implement trauma-informed support strategies and de-escalation interventions (see next theme). Moreover, the likelihood of restraint use being seen as (re)traumatizing is minimal as this documentation pattern paradoxically puts responsibility squarely on the patient.

As previously stated, most psychiatric institutional policies are intended to ensure that restraints are used only after all reasonable efforts are made to find alternative, less coercive, and restrictive measures and identify de-escalation strategies. However, the analysis of chart documentation reveals scant documentation of MHPs considering de-escalation strategies compared to the frequent and detailed documentation of patients' responses that led to restraint use. This asymmetrical documentation pattern, the over-documentation of patients' responses and under-documentation of MHP actions, works in collusion with the inferences of patient passivity to recede MHPs into the psychiatric chart. It serves to erase MHPs' actions leading up to and during restraint use. This is seen in Glenda's chart:

> At 1540 patient was yelling at others, agitated, not redirectable and not willing to follow directions, and threw a book into the nursing station, trying to hit staff. Security was called and patient walked over to the seclusion room by staff and security. At 1545 pt. was put into seclusion room, and is currently being monitored by staff. (nurse)

Similarly, in the excerpt below, the language of "disruption" is used to justify the use of locked seclusion, while the reader is left wondering about the MHP's use of alternative de-escalation strategies before Glenda was mechanically and chemically restrained, "Client has been loud and disruptive most of the night thus far. Singing, talking loudly, and being verbally abusive towards staff. Redirection ineffective" (nurse).

Discounting Structural Violence

The context of gender, race, and class, and structural violence in the form of sexism, racism, colonialism, and classism as they relate to patients' expressions of distress and restraint use are never explicitly documented. However, this analysis of restraint use documentation raises important considerations related to how sexism, racism, colonialism, and classism are implicated in the "unevenness" with which restraints are employed in psychiatric institutions. First, the analysis reveals that men engaging in sexual violence were not as readily subjected to control interventions as were women who were often characterized as "sexually inappropriate." While sexual violence enacted by men was often minimized through euphemistic language such as "sexually inappropriate," women were restrained because they were being "sexually inappropriate." Ebo, who engaged in sexually threatening behavior toward a female co-patient, was documented thus:

> Client was observed entering a female co-client's personal space. Female co-client kept asking her [sic] to move away and client did not listen. Writer then intervened and redirected client. A short while later client was observed touching another female co-client inappropriately on the waste [sic]. Client was again redirected and counselled regarding the need for him to respect people's personal boundaries. (nurse)

> At 1840 hours, staff LT reported he saw client "grabbing" co-client's behind while they were walking in the hallway; co-client did not resist and observed her smiling instead. Will monitor client. (nurse)

Comparatively, Glenda was placed in locked shut down for blowing a kiss to a male staff and, shockingly, because another patient was "sexually disinhibited": "Called to unit at 1130PM to assess this and another patient. She has been pacing the unit and is having difficulties settling.

She is sexually inappropriate towards male staff. Observed to blow a kiss. Difficult to redirect" (psychiatrist). The psychiatrist continued: "In locked seclusion due to her agitation and since there is another patient on the unit who is sexually disinhibited and has been targeting her" (psychiatrist).

Second, in several charts, documentation suggests that the broader context of colonial violence and intergenerational trauma evaded MHPs' understanding of patients' distress responses. For example, notwithstanding Devon's incarceration and trauma history as an Indigenous man, his distress was described as "what appears to be tantrums" (psychiatrist). The psychiatrist identified psychosis or "personality factors" as the possible source of his distress response:

> Code white activated and with security present, Devon accepted prn lox 25mg and ativan 2mg po. Still unable to accept any ownership of his b/h [behavior]. Making statements that the police "beat up" people and now security was there to do the same. Self-entitled with his own needs and unable to appreciate the response his b/h was provoking in the staff. (psychiatrist)

> This is the first adm for this 20 year old with cannabis addiction, a one year hx [history] of increasing paranoia and erratic behaviour and a strong family hx of psychosis, mainly bipolar disorder. Pt has explosive outbursts and what appear to be tantrums. It is hard to distinguish between personality factors and psychosis/mood disorder as the cause of these outbursts. (psychiatrist)

The psychiatrist dismissed Devon's experiences of police violence and the retraumatizing effects of security enforcing coercive measures. Similarly, institutional anti-Black racism and, relatedly, police use of deadly force against Black people were not factored into the MHP's response to concern expressed by Ebo's wife about calling police:

> Met with wife and cousin.
>
> Reviewed hx with them.
>
> Wife reports he has been physically assaultive, has struck her and thrown lighters at her. She fears for her safety but was afraid to call police,

> even though family advised her to do so. I advised her to call the police. (psychiatrist)

The experiences of the Indigenous and Black patients described above are different from Joanna's, a 28-year-old, white, cis woman (heterosexual, income >30K), who was noted to be a professional. A closer look at documentation in Joanna's chart explicates institutional privilege at the intersection of gender, race, and class. Joanna's distress was often documented in terms of being fearful and anxious, rather than aggressive, and situations in which other patients were restrained (e.g., "trying to get out of door") were met with significant de-escalation efforts by MHPs:

> Client was escalating in her behaviour. Yelling trying to get out of door. Writer attempted to close door to prevent another client from entering and client made a lunge for the door. Given Lorazepam 2 mg and Olanzepine 5 mg with a great deal of persuasion. Client is very paranoid and fearful. (nurse)

Joanna's verbal threats toward co-patients were contextualized, with MHPs focusing on her emotional experience as a justification for her behavior:

> Patient approached 2 co-patients on separate occasions while they were being attended to by staff. She yelled at them in a threatening manner as though they posed a threat to her, believing they could cause her some harm. Settled with PRN meds. (nurse)

In striking contrast to other charts, MHPs documented their attempts to understand Joanna's distress, spending time with her in the lounge after she threw a weighted chair rather than locking her in a seclusion room:

> Pt was getting slowly agitated and loud this morning offered prn same refused, able to calm down on her own for a while. She came out of a group this afternoon and started screaming agitated and loud. Went to the lounge area grabbed a chair and threw the chair in the middle of the lounge area as witnessed by staff. Staff approached very agitated unable to

calm down, prn was offered refused initially and took same after encouragement. Pt became tearful still wants discharged from hospital. Given Olanzepine 5mg and Lorazepam 2mg po prn at 1340hrs for agitation.

Pt stated did not want to be in hospital was fine and does not know why she's in hospital. Staff spent time with her in the lounge area. (nurse)

Documentation also suggests implicit understanding that the hospital milieu (15-minute checks, changing staff, having co-patients around clients) may have impacted Joanna's distress responses:

took hs medication and tried to settle to bed

was startled by staff doing rounds and quickly became very agitated–asking if 2 cops were here–asking where her boyfriend went–not recognizing staff/saying we look different–not able to re-settle–not receptive to reassurance/re-direction to room; accepted prn olanzapine 5mg and lorazapam 2mg @ 2230hrs with minimal encouragement–pending effect. (nurse)

patient is suspicious and paranoid, "other patients were walking around my room, they are dangerous, I am afraid of them." reassurance and support given to patient. (nurse)

Code white documentation for Joanna offers more detail about her emotional state ("very scared," "does not feel safe on the unit," "wanting to go home"), although MHPs also used the passive voice (e.g., "was placed on Form 1," "placed on locked seclusion") in documentation to recede into the chart:

Pt came at the nursing station @ 2135 hrs and barge in to the door as soon as staff opened the door to talk to pt. Pt was very anxious, agitated and appeared very scared stated that she does not feel safe on the unit wanting to go home, sat by the window in the nursing station refusing to come out, grabbed the phone and tried to call 911 while staff was trying to talk to her and encouraging to take PRN meds to help her calm down, declined offered med. Pt. escalated, uncooperative, resistive to redirection, grabbed and hit staff on the arm and scratched another staff on her left arm. Code white was called @ 2142hrs. Pt continues to be resistive to redirection while security staff and duty doctor present. Pt.

> continues to refuse oral Prn, was placed on Form 1, stat IM meds ordered. Pt then decided and took oral Prn Lorazepam 2mg po and Olanzapine ODT 5mg @ 2215hrs. Pt placed on locked seclusion and maintained on constant observation for 1 hour. Pt remains awake banging on the door. Locked seclusion with q15 min observation continues. (nurse)

Overall, the restraint-related documentation in Joanna's chart at the intersection of gender, race, and class—middle-class, white woman—might be characterized as more empathic and humane. Documentation suggests that MHPs were less fearful of Joanna as a middle-class, cis, white woman than of other patients who were poor, Black, or Indigenous, and therefore focused more on her emotional experience and did not subject her to coercive measures as frequently.

Concluding Thoughts

This chapter illustrates the ways in which the trauma enacted on patients through the use of restraints is minimized in psychiatric documentation. Chemical, physical, and mechanical restraints are (re)traumatizing for patients. They cause psychological trauma, physical harm, in some instances death, and lack of trust in MHPs and mental health care institutions (Funayama & Takata, 2020; Kontio et al., 2012). Qualitative research with patients who have experienced restraints or seclusion documents significant emotional trauma from this practice. Emotions identified by patients include anger, humiliation, confusion, loneliness, helplessness, and powerlessness (Khatib et al., 2018; Kontio et al., 2012; Spinzy et al., 2018). Many patients with histories of institutional trauma report a retriggering of previous trauma (Khatib et al., 2018; Wong et al., 2020).

Strategies to reduce rather than eliminate restraint use are often discussed in the literature (Kinner et al., 2017). In 2008, the National Association of State Mental Health Program Directors in the United States released "Six Core Strategies to Reduce the Use of Seclusion and Restraint," a report and recommendations that have been enacted throughout the US, showing reductions of 47% to 92% in the use of restraint in 70 facilities (Fernández-Costa et al., 2020; Huckshorn, 2008; Kinner et al., 2017). Common components of restraint reduction strategies in hospitals include enhanced staff training in de-escalation techniques and personalized treatment plans for those who are at risk of being restrained or placed in seclusion (Fernández-Costa et al., 2020; Jacob et al., 2016; Richmond et al., 2012). Studies also indicate that adequate staffing and support are crucial to ensure least-restraint policies (Brickell et al., 2009; McKeown et al., 2019). Other studies have recommended that MHPs receive regular training on the impact of restraints and strategies to minimize them. For example, one study recommended regular educational sessions on the potential psychological impact of restraints as well as reflective practice groups to review and interrogate policies and procedures, while another recommended education and support on strategies to reduce restraint use in specific groups, for example, older people with poor mobility (Huizing et al., 2007; Walker & Tulloch, 2020). Several studies recommend improved communication with patients (e.g., providing information and support in a calm and sensitive way, mandatory staff presence during the duration of restraint use, and supportive debriefing) (Lanthén et al., 2015; Ling et al., 2015).

Qualitative research with patients who have experienced restraints suggests improvements are needed in several domains. The first is in preventing these events, including providing patients with meaningful activities, documenting patients' wishes, and making patient–staff agreements ahead of time (Kontio et al., 2012; Lamanna et al., 2016; Ling et al., 2015). Other recommendations include more comfortable units and the ability to go outside for fresh air and smoking breaks (Kontio et al., 2012; Lamanna et al., 2016; Ling et al., 2015). Supporting patient autonomy by allowing them to make decisions on clothing and when to eat, sleep, and attend to hygiene were also strategies that reduced the frequency of control interventions (Kontio et al., 2012). The second is improved communication and interaction before, during, and after the restraint (Kontio et al., 2012; Ling et al., 2015; Spinzy et al., 2018). Patients have expressed a desire to have MHPs speak with them genuinely and empathically during the restraint or seclusion and also have expressed a need to discuss the event and their feelings afterward in a supportive, empathic, and nonpunitive way (Khatib et al., 2018;

Kontio et al., 2012; Lanthén et al., 2015). Engagement in the treatment process helped patients to feel like active participants in their own care and less confused, frightened, and perplexed (Kontio et al., 2012; Krieger et al., 2018; Lanthén et al., 2015). Third, soothing experiences during restraint use, including reading books, listening to music, or engaging in exercise can be helpful (Kontio et al., 2012; Krieger et al., 2018). Beautiful and comfortable physical spaces were also identified as helpful and healing rather than punitive and frightening, including comfortable bedding and access to a bathroom, proper ventilation, and appropriate temperature (Khatib et al., 2018; Kontio et al., 2012). The fourth and most important improvement is the engagement of patients in the practical development of policies and procedures (Kontio et al., 2012). As Kontio et al. (2012) point out, to ensure high-quality, patient-centered psychiatric care, engagement of and co-creation with patients in policy formation is necessary. The importance of incorporating service users' perspectives in the development of inpatient "aggression" management strategies has been identified by several organizations as being of utmost importance (Kinner et al., 2017; Kontio et al., 2012; Tingleff et al., 2019).

It is important to point out that MHPs working in acute care settings often face very real violence in their day-to-day work, as demonstrated in the documentation in this chapter and in the literature (Lanthén et al., 2015; Walker & Tulloch, 2020). These experiences of physical aggression, sexualized language, and verbal threats can lead to distress and burn out. While violence flows in all directions (toward patients, between patients, and toward staff), it is governed by hierarchal power dynamics that disadvantage patients, particularly patients who are marginalized because of gender, race, sexuality, and class. An interrogation of institutional and structural violence, as discussed in this chapter, is required to create a more humane and equitable system. As an emergency department psychiatrist, I understand that situations will occur in which I will need to make the decision to restrain a patient because I—and my colleagues—believe that the risk for violence is high and know that other strategies have been ineffective. However, I also understand that my decision will cause trauma for a patient. As such, it is my responsibility, along with the institution, to review how institutional policies and procedures contribute to patient distress and aggression, and to support patients who experience this trauma at our hands. Changing documentation patterns to make more visible the trauma done to patients and MHPs' attempts to de-escalate a situation before restraint is used, as well as their and security's active participation and coercion in instances of control interventions are critical steps in this process.

Acknowledgements The author wishes to thank Dr. Andrea Daley and Dr. Merrick Pilling for their analytical contributions to this chapter.

Notes

1. The term "restraint" recalls the act of limiting an individual's freedom of movement. It is, nonetheless, important to emphasize how the term in psychiatry refers more precisely to a coercive act that limits freedom of movement. Any action that is carried out against the will of a patient is considered to be a coercive act (Negroni, 2017).
2. Physical restraints are also referred to as mechanical restraints.
3. A Form 1 is a provision under the *Mental Health Act* in the province where the study took place that allows a physician to detain a patient for a psychiatric assessment for up to 72 hours at a psychiatric facility. A Form 42 (Notice to Person) is always given to a patient to notify them that they are under a Form 1. A Form 1 is only for an assessment and not an involuntary admission per se.
4. A Form 3 (Certificate of Involuntary Admission) is a provision under the *Mental Health Act* in the province where the study took place that is filled out when a patient meets criteria for an involuntary admission.
5. All names given to patients are pseudonyms.

References

Allen, M. H., Carpenter, D., Sheets, J. L., Miccio, S., & Ross, R. (2003). What do consumers say they want and need during a psychiatric emergency? *Journal of Psychiatric Practice, 9*(1), 39–58. https://doi.org/10.1097/00131746-200301000-00005

Brickell, T., Nicholls, T., Procyshyn, R., McLean, C., Dempster, R., Lavoie, J., Sahlstrom, K., Tomita, T., & Wang, E. (2009). *Patient safety in mental health*. BC Mental Health and Addictions Services. https://www.patientsafetyinstitute.ca/en/toolsResources/Research/commissionedResearch/mentalHealthAndPatientSafety/Documents/MentalHealth%20Paper.pdf

Emmanuel, L., Taylor, L., Hain, A., Combes, J., Hatlie, M., Karsh, B., Lau, D., Shalowitz, J., Shaw, T., & Walton, M. (2013). *Module 13d: mental health care: seclusion and restraint: When all else fails*. The Patient Safety Education Program. https://www.patientsafetyinstitute.ca/en/education/PatientSafetyEducationProgram/PatientSafetyEducationCurriculum/Documents/Module%2013d%20Seclusion%20and%20Restraint.pdf

Fernández-Costa, D., Gómez-Salgado, J., Fagundo-Rivera, J., Martín-Pereira, J., Prieto-Callejero, B., & García-Iglesias, J. J. (2020). Alternatives to the use of mechanical restraints in the management of agitation or aggressions of psychiatric patients: A scoping review. *Journal of Clinical Medicine, 9*(9), 2791. https://doi.org/10.3390/jcm9092791

Funayama, M., & Takata, T. (2020). Psychiatric inpatients subjected to physical restraint have a higher risk of deep vein thrombosis and aspiration pneumonia. *General Hospital Psychiatry, 62*, 1–5. https://doi.org/10.1016/j.genhosppsych.2019.11.003

Gaynes, B. N., Brown, C. L., Lux, L. J., Brownley, K. A., Van Dorn, R. A., Edlund, M. J., Coker-Schwimmer, E., Palmieri Weber, R., Sheitman, B., Zarzar, T., Viswanathan, M., & Lohr, K. N. (2017). Preventing and de-escalating aggressive behavior among adult psychiatric patients: A systematic review of the evidence. *Psychiatric Services, 68*(8), 819–831. https://doi.org/10.1176/appi.ps.201600314

Huckshorn, K. (2008). *Six core strategies for reducing seclusion and restraint use*. NASMHPD. http://www.nasmhpd.org/sites/default/files/Consolidated%20Six%20Core%20Strategies%20Document.pdf

Huizing, A. R., Hamers, J. P. H., de Jonge, J., Candel, M., & Berger, M. P. F. (2007). Organisational determinants of the use of physical restraints: A multilevel approach. *Social Science and Medicine, 65*(5), 924–933. https://doi.org/10.1016/j.socscimed.2007.04.030

Itzhaki, M., Bluvstein, I., Peles Bortz, A., Kostistky, H., Bar Noy, D., Filshtinsky, V., & Theilla, M. (2018). Mental health nurse's exposure to workplace violence leads to job stress, which leads to reduced professional quality of life. *Frontiers in Psychiatry, 9*(59). https://doi.org/10.3389/fpsyt.2018.00059

Jacob, J. D., Holmes, D., Rioux, D., & Corneau, P. (2018). Assessing "insight," determining agency and autonomy: Implicating social identities. In J. Kilty & E. Dej (Eds.), *Containing madness: Gender and "psy" in institutional contexts* (pp. 93–118). Palgrave Macmillan. https://doi.org/10.1007/978-3-319-89749-3

Jacob, T., Sahu, G., Frankel, V., Homel, P., Berman, B., & McAfee, S. (2016). Patterns of restraint utilization in a community hospital's psychiatric inpatient units. *Psychiatric Quarterly, 87*(1), 31–48. https://doi.org/10.1007/s11126-015-9353-7

Khatib, A., Ibrahim, M., & Roe, D. (2018). Re-building trust after physical restraint during involuntary psychiatric hospitalization. *Archives of Psychiatric Nursing, 32*(3), 457–461. https://doi.org/10.1016/j.apnu.2018.01.003

Kinner, S. A., Harvey, C., Hamilton, B., Brophy, L., Roper, C., McSherry, B., & Young, J. T. (2017). Attitudes towards seclusion and restraint in mental health settings: Findings from a large, community-based survey of consumers, carers and mental health professionals. *Epidemiology and Psychiatric Sciences, 26*(5), 535–544. https://doi.org/10.1017/S2045796016000585

Knight, S., Jarvis, G. E., Ryder, A. G., Lashley, M., & Rousseau, C. (2021). Ethnoracial differences in coercive referral and intervention among patients with first-episode psychosis. *Psychiatric Services*. appi-ps. https://doi.org/10.1176/appi.ps.202000715

Kontio, R., Joffe, G., Putkonen, H., Kuosmanen, L., Hane, K., Holi, M., & Välimäki, M. (2012). Seclusion and restraint in psychiatry: Patients' experiences and practical suggestions on how to improve practices and use alternatives. *Perspectives in Psychiatric Care, 48*(1), 16–24. https://doi.org/10.1111/j.1744-6163.2010.00301.x

Krieger, E., Moritz, S., Weil, R., & Nagel, M. (2018). Patients' attitudes towards and acceptance of coercion in psychiatry. *Psychiatry Research, 260*, 478–485 (December 2017). https://doi.org/10.1016/j.psychres.2017.12.029

Lamanna, D., Ninkovic, D., Vijayaratnam, V., Balderson, K., Spivak, H., Brook, S., & Robertson, D. (2016). Aggression in psychiatric hospitalizations: A qualitative study of patient and provider perspectives. *Journal of Mental Health, 25*(6), 536–542. https://doi.org/10.1080/09638237.2016.1207222

Lanthén, K., Rask, M., & Sunnqvist, C. (2015). Psychiatric patients experiences with mechanical restraints: An interview study. *Psychiatry Journal, 2015*, 1–8. https://doi.org/10.1155/2015/748392

Ling, S., Cleverley, K., & Perivolaris, A. (2015). Understanding mental health service user experiences of restraint through debriefing: A qualitative analysis. *Canadian Journal of Psychiatry, 60*(9), 386–392. https://doi.org/10.1177/070674371506000903

Lu, W., Mueser, K. T., Rosenberg, S. D., Yanos, P. T., & Mahmoud, N. (2017, July). Posttraumatic reactions to psychosis: A qualitative analysis. *Frontiers in Psychiatry, 8*, 4–6. https://doi.org/10.3389/fpsyt.2017.00129

McKeown, M., Thomson, G., Scholes, A., Jones, F., Baker, J., Downe, S., Price, O., Greenwood, P., Whittington, R., & Duxbury, J. (2019). "Catching your tail and firefighting": The impact of staffing levels on restraint minimization efforts. *Journal of Psychiatric and Mental Health Nursing, 26*(5–6), 131–141. https://doi.org/10.1111/jpm.12532

Negroni, A. A. (2017). On the concept of restraint in psychiatry. *European Journal of Psychiatry, 31*(3), 99–104. https://doi.org/10.1016/j.ejpsy.2017.05.001

Pariseau-Legault, P., Vallée-Ouimet, S., Goulet, M. H., & Jacob, J. D. (2019). Nurses' perspectives on human rights when coercion is used in psychiatry: A systematic review protocol of qualitative evidence. *Systematic Reviews, 8*(1), 1–7. https://doi.org/10.1186/s13643-019-1224-0

Richmond, J. S., Berlin, J. S., Fishkind, A. B., Holloman, G. H., Zeller, S. L., Wilson, M. P., Rifai, M. A., & Ng, A. T. (2012). Verbal de-escalation of the agitated patient: Consensus statement of the American Association for Emergency Psychiatry Project BETA De-escalation Workgroup. *Western Journal of Emergency Medicine, 13*(1), 17–25. https://doi.org/10.5811/westjem.2011.9.6864

Schnitzer, K., Merideth, F., Macias-Konstantopoulos, W., Hayden, D., Shtasel, D., & Bird, S. (2020). Disparities in care: The role of race on the utilization of physical restraints in the emergency setting. *Academic Emergency Medicine, 27*(10), 943–950. https://doi.org/10.1111/acem.14092

Spinzy, Y., Maree, S., Segev, A., & Cohen-Rappaport, G. (2018). Listening to the patient perspective: Psychiatric inpatients' attitudes towards physical restraint. *Psychiatric Quarterly, 89*(3), 691–696. https://doi.org/10.1007/s11126-018-9565-8

Tingleff, E. B., Hounsgaard, L., Bradley, S. K., & Gildberg, F. A. (2019). Forensic psychiatric patients' perceptions of situations associated with mechanical restraint: A qualitative interview study. *International Journal of Mental Health Nursing, 28*(2), 468–479. https://doi.org/10.1111/inm.12549

Walker, H., & Tulloch, L. (2020, May). A "necessary evil": Staff perspectives of soft restraint kit use in a high-security hospital. *Frontiers in Psychiatry, 11*, 1–11. https://doi.org/10.3389/fpsyt.2020.00357

Wong, A. H., Ray, J. M., Rosenberg, A., Crispino, L., Parker, J., McVaney, C., Iennaco, J. D., Bernstein, S. L., & Pavlo, A. J. (2020). Experiences of individuals who were physically restrained in the emergency department. *JAMA Network Open, 3*(1), e1919381. https://doi.org/10.1001/jamanetworkopen.2019.19381

6

Sexual Violence and Psychosis: Intersections of Rape Culture, Sanism, and Anti-Black Sanism in Psychiatric Inpatient Chart Documentation

Merrick D. Pilling

Introduction

The widespread existence and acceptance of sexual violence is supported by what feminist activists and scholars have named "rape culture." Rape culture is the normalization of violence against women and trans people in which sexual violence, in particular, is normative rather than exceptional (Powell & Henry, 2014). Some of the hallmarks of rape culture include the refusal to believe survivors of sexual violence, victim blaming, the sexualization of violence, the objectification of women, and misogyny (Sweeney, 2020). Rape culture is also imbricated with other forms of oppression including misogynoir and transmisogynoir, leaving Black, cis women, and trans people at greater risk of violence,

M. D. Pilling (✉)
Women's and Gender Studies in the School of Social Work,
University of Windsor, Windsor, ON, Canada
e-mail: Merrick.Pilling@uwindsor.ca

© The Author(s), under exclusive license to Springer Nature Switzerland AG 2021

A. Daley and M. D. Pilling (eds.), *Interrogating Psychiatric Narratives of Madness*,
https://doi.org/10.1007/978-3-030-83692-4_6

(re)victimization, and being blamed for the violence they experience (Donovan & Williams, 2002; Preston, 2020; Slatton & Richard, 2020).

In this chapter, I explore how rape culture informs the ways in which mental health professionals (MHPs) document sexual violence in the charts of patients diagnosed with a "psychotic illness."[1] Where relevant, I highlight intersections with sanism and anti-Black sanism (Meerai et al., 2016).[2] I begin by discussing intersections between rape culture, sanism, and anti-Black sanism. I then explore five themes emerging from the data: (1) *normalization of sexual violence*, (2) *sexual violence reconfigured as delusion*, (3) *pathologizing the impact of sexual violence*, (4) *what about the perpetrators? Invisibilizing acts of sexual violence*, and (5) *sexual violence as a symptom of psychosis*. I conclude with a discussion of the institution's responsibility to address sexual violence within and outside of its walls.

Methods

As described in more detail in the Introduction, two research team members (MP and AD) read and analyzed 161 charts for content related to gender, sexuality, race, and class. We devised a codebook that was used to identify common themes across the charts. Based on the coded data, we created memos that refined and expanded key themes. For the purpose of this chapter, I make use of the memos based on data coded under four interrelated themes: representations of violence, representations of trauma, violence on unit, and representations of delusions and/or paranoia. In these memos, I focus on the charts of those who experienced ($n = 20$) or perpetrated ($n = 9$) sexual violence on or off the unit who were also diagnosed with a psychotic illness. I focus on these charts because sanism and anti-Black sanism are highly evident in the charts of those deemed psychotic because of the ways in which such patients' credibility is challenged. Further, these charts are especially relevant to analyses of sexual violence because psychosis has been shown to be associated with experiences of childhood sexual abuse (Bentall et al., 2014; Read et al., 2014). The 20 charts that contain documentation of experiences of sexual violence include five Black, cis women, one

Black, transgender woman, seven white, cis women, one white, transgender woman, two East Asian women, two white, cis men, one white, transgender man, and one Middle Eastern man. The nine charts that contain documentation regarding perpetration of sexual violence include five white men, three Black men, and one Middle Eastern man.

Rape Culture, Sanism, and Anti-Black Sanism

Powell and Henry (2014) define rape culture as "social, cultural and structural discourses and practices in which sexual violence is tolerated, accepted, eroticised, minimised and trivialised" (p. 2). A defining characteristic of rape culture is the frequent and widespread disbelief of survivors of sexual violence. This is because rape culture governs what gets commonly understood and recognized as sexual violence. As Alcoff (2018) explains,

> Rape cultures produce a discursive formation in which the intelligibility of claims is organized not by logical argument or evidence, but by frames that set out who can be victimized, who can be accused, which are plausible narratives, and in what contexts rape may be spoken about, even in private spaces. (p. 2)

The credibility of any survivor of sexual violence is scrutinized. However, the frameworks that dictate "the intelligibility of claims" are racialized. For example, the confluence of anti-Black racism and misogyny, also known as misogynoir (Bailey & Trudy, 2018), means that Black women are even less likely to be believed and more likely to be blamed for the violence they experience (Donovan & Williams, 2002; Onwuachi-Willig, 2018). This is rooted in histories of colonization and enslavement that constructed Black women as sexually deviant and available for sexual violation (Benn-John, 2016; Donovan & Williams, 2002; McGuire, 2010). Likewise, structural transmisogynoir results in the sexual victimization and murder of Black, trans women and transfeminine people and undermines their credibility as survivors of violence (Preston, 2020). Class also plays a role in that those living in poverty are more at risk of sexual violence and being blamed for the violence and poverty they experience (Greco & Dawgert, 2007).

In the context of a psychiatric hospital, rape culture intersects with sanism and anti-Black sanism in problematic ways. Broadly described, sanism is a form of oppression affecting Mad people and those with mental health diagnoses. Sanism is normalized within society at large and within the mental health system as "good clinical practice" (Abdillahi et al., 2017, p. 109). Sanism positions Mad people as having "a flawed or disordered way of seeing, perceiving, judging, and thus, knowing reality" (Liegghio, 2013, p. 126). This is particularly salient for those deemed psychotic, as this state of being is associated with phenomena such as unusual beliefs ("delusions") and hearing voices ("auditory hallucinations"), which further challenges the nature of a shared reality between patient and practitioner. As will be discussed, sanism compounds rape culture in that sexual violence is perhaps even less likely to be recognized as such when experienced by those whose realities are already constructed as questionable.

Conceptualizations of sanism must also consider race and racism, because "sanism exists on a continuum depending on privilege, and it is always and especially compounded when it is visited on racialized bodies" (Meerai et al., 2016, p. 22). Anti-Black sanism is rooted in psychiatry's history as a tool of colonization and a means of upholding slavery (Abdillahi et al., 2017; Fernando, 2010; Tam, 2013). Abdillahi et al. (2017) argue that Black and African Canadians experience anti-Black sanism in mental health services related to "communication, diagnosis, hospitalization, treatment, intervention, and the involvement of the criminal justice system" (p. 122). This is evidenced by various phenomena including the disproportionate numbers of Black people in the mental health system, high numbers of involuntary hospitalizations, and the disproportionate diagnosing of Black people with schizophrenia (Abdillahi et al., 2017; Fernando, 2010; Metzl, 2010).[3] Anti-Black sanism means that Black women, in particular, are subject to stereotypes that position them as "hypersexual, 'sick,' unsafe or dangerous, and inappropriate" (Meerai et al., 2016, p. 24). When compounded with rape culture, these anti-Black sanist stereotypes create a toxic mix.

Incidents of sexual violence are often characterized as being a matter of one party's word against the other's and this characterization hinges on problematic understandings of credibility and innocence imbued with structural power dynamics based on race, gender, madness, class, sexuality, and disability. This chapter considers documentation about patient histories of sexual violence as well as incidents of sexual violence occurring on inpatient units during patients' institutionalization. It is, therefore, important to note that in a psychiatric hospital, there is an immense amount of surveillance and documentation regarding the actions of patients on the unit (including, for example, what a patient eats, where they are on the unit and who they talk to, when and for how long they sleep, and sometimes even whether they have had a bowel movement). In some cases (such as the risk of harm to self or others), patients may be placed on "Q15," which means that a nurse is supposed to observe the patient every 15 minutes and record this in the chart, or on "constant observation," meaning that they are being continuously watched. It is, therefore, the kind of environment where the type of evidence of sexual harassment or assault that is usually impossible to provide does, or could, exist. And yet, as will be shown, this does not shape the outcome in favor of the survivor. This is the power of rape culture: it shapes how practitioners perceive, understand, and document patient experiences of sexual violence in ways that normalize and invisibilize sexual violence. As will be shown, this is further compounded by sanism and anti-Black sanism.

Normalization of Sexual Violence

Rape culture means that sexual violence is mundane, or as Sweeney (2020) states, "within a rape culture, sexual violence is routine and predictable, not extraordinary or aberrant" (p. 289). The operation of rape culture to normalize sexual violence is evident in the charts, which contain numerous examples of women who experienced sexual harassment from other patients on the units, though the words "sexual harassment" or even "harassment" are rarely used. For example, the chart of Ama,[4] a Black African, heterosexual, cis woman, contains evidence that she experienced sexual harassment from another patient on the unit. The first mention of this was two days after admission, when a nurse documented the following:

> Client calm, and appropriate. Spent time watching TV in lounge. Does not interact with co-clients.
>
> Compliant with meds, and ate snack.
>
> Reports feeling a restless feeling at night and has requested to switch rooms in the morning if possible as she feels it is related to her room. Denies SI/HI [suicidal ideation/homicidal ideation].

This note would be unremarkable if not for documentation written two days later by a psychiatrist, which provides more context: "Pt slept poorly last night because co patient [name redacted] kept peering into her room, making her feel unsafe. So she was napping this afternoon. She is using her off ward passes and is appreciative of this." In light of the information provided about another patient "peering into her room," the nurse's note two days previous that Ama was "restless" and requested to change rooms becomes more clear, as does the nurse's vague description that the issue is "related to her room." This vague wording and the lack of accompanying context leaves Ama's request to change rooms and the reason for the request open to (mis)interpretation as seemingly bizarre and potentially delusional, perhaps especially so when followed by "denies SI/HI" and in light of her diagnosis of a psychotic illness.

There is no further description of the incident(s) in the psychiatrist's documentation or indication of action taken to address the situation. The chart excerpt reads as if the incident was mentioned by the psychiatrist in passing and was only seen as relevant to explain Ama's afternoon nap. There is no indication that this is seen as harassment or any expressed concern with Ama's safety on the unit.

Twelve days later, a social worker documented Ama's experiences of harassment in more detail:

> She reports feeling extremely uncomfortable on [unit number redacted] as a Nigerian male patient on the unit is constantly at her door and opening the curtain on her door to look through the window at her. She has advised her nurse and a towel has been placed over her window and

> her room was moved so that this man was not near her room, she reports but he continues to approach her when she is out of her room and she has told him twice to leave her alone but he has continued to approach her so she is now spending her time in her room as she feels too uncomfortable to be out of her room because of him she reports.

It is noteworthy that there is no documentation from nursing regarding the harassment or the "solution" of changing Ama's room and placing a towel over her window. It is possible that the social worker chose to document these experiences of harassment in more detail as a (limited) form of resistance to the normalization of violence within the institution. However, the veracity of the patient's narrative is tempered by the repeated use of "she reports." The choice to name the nationality of the male patient is conspicuous and raises questions about why this information was included and whether anti-Black sanism was at play. Was the social worker's use of "Nigerian" coded language for "Black"? Given that Black Africans, in particular, are constructed as "uncivilized" (Kumsa et al., 2014), and Black men are more likely to be identified as perpetrators of sexual assault (Miller, 2019), did the social worker find the patient's narrative worthy of documentation because the perpetrator was, presumably, a Black man? The chart contains no further documentation of these incidents of harassment by any practitioner, even though Ama was institutionalized for an additional 33 days. The words "sexual harassment," "harassment," and "stalking" are never used, and there is no indication that any further action was taken to protect Ama and make her feel safe on the unit. There is also no expressed consideration that Ama's concern for her own safety might have a direct impact on her mental distress.

Ama's case is not an isolated incident; there are other examples of documentation that suggest the influence of rape culture in the normalization of sexual violence within the institution. In another example, the chart of Breanne, a white, heterosexual, cis woman includes documentation by a social worker regarding a meeting with Breanne's son. The social worker wrote that Breanne's son "expressed some dissatisfaction" with two "incidents" that took place during his mother's previous admission, including "a male co-patient entering client's room and another involving a co-patient trying to hit client. Said the staff acted as if 'nothing happened'." Additionally, the same chart includes another entry from a social worker who noted the following:

> Client mentioned that she had found it upsetting watching 2 female co-patients "fighting" and that she found it scary. Also mentioned that there has been a different female co-patient that is sometimes intrusive and tries to touch her. She said nurses are aware of this. Writer to discuss with them as well.

Again, the social worker may have been trying to resist the normalization of violence in the institution by choosing to document this information and by recording the intention to discuss the matter with the nursing staff. Similar to Ama's chart, there is no documentation from nursing regarding the "intrusive" co-patient despite the fact that the social worker stated that nursing was made aware of the problem, suggesting that practitioners make decisions about whether to include this kind of information in the chart. However, the word "intrusive" as opposed to "harassment" is notable. Given the likelihood that practitioners do not recognize the potential of sexual violence perpetrated between women (Daley & Ross, 2018) was the incident less visible as harassment because the perpetrator was a woman? There is no indication that these incidents were taken up by other practitioners as playing a role in Breanne's distress, and there is no further documentation of these incidents in the chart.

In Ama's and Breanne's charts, their experiences of sexual harassment seem to be understood as a mundane and mostly unremarkable part of life in a psychiatric hospital. The incidents were perhaps seen as a nuisance to be managed but not as sexual harassment. This is one way in which rape culture operates within the psychiatric institution: experiences of sexual violence are normalized. In the examples I have discussed thus far, it is not clear if (or how) the women's diagnoses of a psychotic illness played a role in how their reports of violence were taken up and responded to by practitioners, though one may assume sanism played an implicit role. In the next section, the way in which sanism and anti-Black sanism compound rape culture becomes more explicit.

Sexual Violence Reconfigured as Delusion

Based on the data from the pilot study preceding the current study, Daley et al. (2011) question why violence against women is not featured in the charts as a social issue. They suggest that practitioners may reconfigure sexual violence and trauma as illness, thereby positioning it as a biomedical issue rather than a social one (pp. 9–10). The data from the current study supports this contention. Specifically, experiences of sexual harassment and violence are reconfigured as delusional thinking. For example, in the chart of Carly, a white, heterosexual, cis woman, a psychiatrist documented the following:

> She was initially quite calm but became more irritable. She recounted the events of the previous evening and said that she had been bothered by some male clients on the ward who were making noise and peered into her room. She said that she felt threatened.

In the same note, the psychiatrist went on to state:

> I could not ascertain whether her thoughts about co-patients were rooted in paranoia or represented an accurate recollection. She reported sleeping poorly with increased goal-directed activity and racing thoughts. She was offered PRN medication but declined. She said that she feels like she's being drugged. The patient seemed more re-directable than yesterday. She agreed that if nursing staff had concerns about her level of agitation outside her room, she would take PRNs then.

The psychiatrist characterized Carly as "irritable," indicating a disregard for the emotional impact of being harassed, feeling threatened, and not being believed. It stands to reason that someone who feels threatened and unsafe may have trouble sleeping. However, this was pathologized by the psychiatrist as "sleeping poorly with increased goal-directed activity and racing thoughts." The psychiatrist's solution was to offer medication, a biomedical solution to a social problem. The patient's resistance to this offer ("she said that she feels like she's being drugged") is a powerful and disturbing indication of the harmful power dynamics of the institution, which can (and often does) coercively drug patients. The threat of coercion is present in the psychiatrist's veiled warning about taking medication if the nurses "had concerns about her level of agitation." There was no expressed concern for Carly's safety or how the harassment could impact her mental distress. Instead, Carly was characterized as a problem to be managed, and the incidents of harassment were reconfigured as paranoid delusions. This was reinforced and stated more strongly in the discharge note, where a psychiatrist referred to Carly as having "paranoid delusions that she was unsafe on the unit." Given the frequency with which similar incidents of harassment are mentioned across the charts (with many more likely undocumented) and the widespread existence of sexual violence in society, this refusal to believe Carly and reconfigure her experiences as delusional indicates the influence of rape culture and sanism on the practitioners' judgment. A hallmark of rape culture is the disbelief of survivors (Powell & Henry, 2014; Sweeney, 2020). In the context of the psychiatric institution, this was compounded by sanism, in which Carly was not believed because her experiences of sexual violence were construed as a symptom of psychosis (delusions). However, problematic this may be, the tone of Carly's chart reflects relative agency and "respect" for her as a middle-class, white woman (see also Chapter 3, *Recognizable Genders = Credible Narrators*). This becomes more evident when contrasted with an example from another chart, which brings the role of anti-Black sanism to the fore.

A strikingly egregious example of rape culture compounded by anti-Black sanism is found in the chart of Vea, a heterosexual, cis woman who was noted at some points to be mixed heritage, at others to be Indian-Caribbean, and in other places as Black-Caribbean. On day 51 of her admission, the chart records that Vea told a nurse that she had been sexually assaulted by another patient on the unit. A physician subsequently documented salient details about the assault including the initials and medical record number (MRN) of the person she was assaulted by, that he had a history of sexual assault, and that Vea had reported him entering her room several nights in a row when she was sleeping and heavily sedated:

> Pt very upset, and says she was sexually assaulted several times on ward over past week. She says she found large amounts of "blood and ejaculate" in her vagina and rectum and semen and blood stains on her underwear. She does not recall this happening but says she has found co-patient [initials and medical record number redacted] (who does have a hx [history] of sexually assaultive behaviour) in her room at night 4 x the past week when she was half awake and believes the assaults happened when she was asleep.

Several hours later, the patient was interviewed by police, then taken to another hospital where a rape kit was completed. She was given antiviral medication and an emergency contraceptive. She was returned to the hospital that same night just before midnight and transferred to another unit. Very early the next morning, she was seen by a psychiatrist who questioned the legitimacy of her narrative regarding the assault. An excerpt from this psychiatrist's note reads as follows:

> - I asked if it was possible that an assault didn't happen and she became furious at this suggestion, insisting to know who was spreading the rumour that she has claimed sexual assault several times and to know who would challenge her
> - she said she felt like she was being interrogated
> - I settled her and indicated I just wanted to be sure the antivirals being given to her are appropriate and that there is a reasonable chance that she was assaulted as there is no clear evidence that this occurred;
> - she again was adamant that she had vaginal tearing (no evidence of this from ER note, not clear if internal exam done), and knew there was sperm inside her

At the end of this note, the psychiatrist completed a mental status exam, which reads:

> Approp[iately] dressed, no pain; met patient with nurse; initially calm but then became very defensive and then verbally aggressive; no voiced SI/HI; very sure that she was assaulted sexually despite having no image or recollection or suspicion who did this on the unit; insight—limited; judgment—no imminent risk.

Following a grueling night that included Vea's expressed emotional distress and a rape kit (an experience that many likely find intrusive and emotionally difficult), Vea was questioned as to the veracity of her account. It is no wonder that she "bec[a]me furious." When Vea said that she felt as if she was "being interrogated," the psychiatrist made a move to innocence, gaslighting the patient by characterizing his questions as ensuring the antiviral medication was "appropriate," while yet again denying the assault ("no clear evidence that this occurred"). The psychiatrist reaffirmed the denial in stating that there was "no evidence" of vaginal tearing, despite the fact that it was "not clear if internal exam done." Vea's response to having her experiences blatantly denied was further characterized as "defensive" and "verbally aggressive" in the mental status exam. Notably, the information about the perpetrator, including initials and MRN number recorded earlier in the chart, was not referenced in this psychiatrist's note, and he instead stated that Vea has "no image or recollection or suspicion who did this on the unit." In the same note, the psychiatrist further stated that it was "unclear if sexual assault on [name of unit redacted] actually occurred– staff documented that male patient may have been in her room." This contradicts earlier documentation that contained details about the perpetrator including that he had a history of committing sexual assault. The psychiatrist also suggested transferring Vea back to the unit where the assault occurred, where presumably, the perpetrator was still staying. A mental status exam completed later that day by a nurse likewise ignored these salient details: "Does not want to return to [unit number redacted] because a man came into her room two nights ago and sexually assaulted her. She did not see his face. No idea of who he was. Mood is stable. No suicidal thoughts."

Three days later Vea was assessed by a psychiatrist who further questioned the legitimacy of her reports and more clearly reconfigured Vea's account as delusional:

> We reviewed her history. She was transferred from [unit number redacted] as she alleged that she was raped by a copatient. I believe from the notes that this is certainly questionable. This evening, knowing her history, I asked RN to join the meeting. She was very sexually preoccupied, rambled on and on about people raping her, others being raped, her

> family [description of sexual violence redacted],who have also been raping her frequently. She gave very graphic descriptions about [description of her experience following the assault redacted]. She provided inconsistent reports of her past psychiatric history, and hopefully we can connect with her family at some point when she is a bit better.
>
> Mental Status Exam: Clearly delusional on some topics, and questionable on others. Ranted on and on, and was difficult to interrupt at times. Cried uncontrollably at one point, largely out of the blue, when she started talking about everyone raping her and the police not doing anything about it. She was upset that I did not act on her reports of rape, and I asked what she wanted to me to do, and she replied that the College of Physicians should have taught this to me.

The psychiatrist's documentation indicates a selective reliance on existing documentation ("I believe from the notes") rather than on Vea's account of her experience. In yet another classic indication of the influence of rape culture, the psychiatrist suggested that he was the one who needed protection, presumably from false allegations of sexual assault ("knowing her history") and asked a nurse to join the meeting (an inversion of the actual power dynamic). The psychiatrist pathologized the patient's account as "sexual preoccupation" and demonstrated a callous indifference to her experiences ("rambled on and on about people raping her"). The "graphic" information noted about Vea's experience of the assault was consistently noted by various practitioners following the assault, but this was not perceived as evidence of consistency by the psychiatrist, who cast further doubt on her credibility by noting vaguely that she "provided inconsistent reports of her past psychiatric history." The mental status exam that concludes the note indicates that the psychiatrist believed the experience of assault was likely a delusion, thereby reconfiguring experiences of sexual violence as illness. The mental status exam also further demonstrated the psychiatrist's brutal dismissal of the patient; he saw Vea's emotional response of "crying uncontrollably" as being "out of the blue" when discussing experiences of rape and police inaction. Would the psychiatrist's response to the tears of a middle-class, white, cis woman have been the same? Anti-Black sanism is at work to deny the humanity of Vea, a homeless woman who was starkly and repeatedly characterized as an "obese Black female" throughout her chart.[5] Vea's diagnosis of a psychotic illness and the fact that she was already deemed delusional prior to the assault compounds anti-Black sanism with rape culture to render her not credible and unworthy of compassion and respect.

Pathologizing the Impact of Sexual Violence

In the examples discussed thus far, I have explored how rape culture, sanism, and anti-Black racism shape practitioners' documentation about sexual violence occurring on inpatient units, arguing that it is normalized and that experiences of sexual violence on the unit are sometimes pathologized and reconfigured as delusional thinking. In what follows, I discuss how rape culture also shapes practitioners' documentation about patient histories of sexual violence by failing to meaningfully connect past experiences of abuse to current distress and by pathologizing the impact of abuse as indicative of psychotic illness. This section further illustrates the points made above regarding the normalization and reconfiguration of sexual violence, while delineating how the impact of past violence is also pathologized. In other words, in addition to incidents of violence being doubted or denied through normalization and pathologization, the ways in which the impact of violence manifests itself in patient behavior is likewise pathologized as illness. This further demonstrates the confluence of rape culture with sanism in that sexual violence is invisibilized through the pathologizing of survivors' realities post abuse.

It is well established that many psychiatric service users are survivors of childhood sexual abuse (Bentall et al., 2014; Herman 2015; Read et al., 2014). Herman (2015) writes that those who have experienced childhood sexual or physical abuse often present with "a bewildering array of symptoms" and are frequently "misdiagnosed and mistreated in the mental health system" (p. 123). People who experience childhood abuse often develop behaviors and strategies that enable their survival in abusive environments, but which become "maladaptive" in adulthood and are construed as mental disorders (Herman, 2015, p. 127). Many are given numerous psychiatric diagnoses instead of contextualizing "symptoms" as a response to abuse. Scholars have argued that psychosis, in particular, should be reimagined as an understandable reaction to abuse, trauma, and adverse life events rather than as an individualized, biochemical disorder (Johnstone, 2011; Romme & Escher, 2012; Unger, 2019).[6]

This phenomenon is evident in the chart of Asher, a middle-class, cis, heterosexual man who was noted by various practitioners as Middle Eastern, but who was described by the psychiatrist on admission to have "appeared as Caucasian man," indicating that he was perceived as white. A nurse initially charted "Trauma History Identified: No" in the triage note. However, the admission note written by a psychiatrist the next day indicates vaguely "history of trauma @14y." Three days later, a psychiatrist indicated that Asher intended to start attending a support group for male survivors of sexual abuse. That same day, a social worker noted in passing, "Writer and patient discussed concerns about self-worth, trauma, and persistent symptoms of psychosis." There is no other mention of sexual abuse or trauma during Asher's 10-day admission aside from a verbatim copy-and-paste section of the discharge note that includes the phrase "history of trauma @14y" described above.[7]

Asher's experiences of childhood sexual abuse were barely noted at all and were never meaningfully connected to his distress. Instead, the impact of abuse was pathologized. The following documentation written by a psychiatrist is telling in this respect:

> Today an anxiety screen was performed. At baseline patient tends to over worry, he has difficulty sleeping and feels on edge. He also tends to be concerned about how others perceive him. In the past he has had obsessive thinking about inadvertently harming others and tends to profusely apologize when there is no reason for this.

Those who are abused as children often believe that they are to blame and that the abuse happened because there is something wrong with them (Herman, 2015, p. 105), and as adults, they continue to feel culpable. This feeling of culpability can manifest as chronic apologizing (Kippert, 2016).

The day before Asher was discharged, a psychiatrist wrote, "Given his longstanding obsessive worrying about how others might perceive him, we reviewed that the use of SSRI's would be indicated. We discussed risks and benefits of Zoloft." In the same note, the psychiatrist completed a mental status exam in which they noted:

> Affect is brighter. He is visibly calmer. He is able to challenge his paranoia as irrational thoughts. He also continues to worry if he might have inadvertently hurt someone and this worry seems obsessive in nature. No SI [suicidal ideation] or violent ideation. Insight and judgment are good.

The psychiatrist diagnosed Asher with a disorder with "psychotic features," started him on medication, and discharged him the next day. In this example, the psychiatrist's focus was on biochemical factors and solutions. Asher's experiences of distress (worrying about being a predator or being perceived as one) and his tendency to apologize make sense within the context of his trauma history of childhood sexual abuse, but this was never contextualized in the chart. Instead, these thoughts were characterized as "paranoia" and "obsessive," and the psychiatrist's focus was on diagnosis and using medications to quell the distressing thoughts. Despite the fact that Asher also received a referral for cognitive behavioral therapy to change what got called his "obsessional thinking," these "obsessions" were clearly seen as meaningless (or generated by biochemical factors) in that they were never explicitly connected to his experiences of childhood sexual abuse. Thus, the impact of sexual violence was reconfigured as pathology. Once again, this shows evidence of rape culture compounded with sanism, in which sexual violence is erased by attributing the impact of abuse to biomedical factors. Notably, the tone of the chart and the agency accorded to Asher reflect his social location as a white-passing, middle-class, cis man. This becomes more evident when contrasted with an example from another chart, which highlights the role of anti-Black sanism. Asher's chart shows marked respect for his agency in contrast with that of Irie, a homeless, Black Caribbean, transgender woman.

Similar to Asher, the impact of sexual violence was barely noted in Irie's chart and was never meaningfully connected to her experience.

Likewise, the impact of violence was pathologized in that the way it manifested in her behavior was not recognized. On intake, Irie was charted as having experienced "physical abuse as a child": "Trauma History Identified: Yes; Trauma History Comments: Reports physical abuse as a child." On day three of her admission, a nurse documented that Irie "started to become agitated, speech-loud, stated toward staff—'hey look! I am abused all my life! My father is an abuser!'" Later the same day, another nurse documented the following:

> Upset about this hospitalization and blames his [sic][8] father for this. Claims he's [sic] "not sick. Has a place to stay outside a bank. Father and brother sexually molested me." Believes he's [sic] the ruler of the world. He's [sic] Rihanna, Sailor Moon. He's [sic] creative. "You know I'm transgendered. I've used several female names before."

In this excerpt, Irie's account of sexual violence and her transgender identity are presented alongside fantastical beliefs such as being Rihanna and Sailor Moon, suggesting that the nurse may have seen all these assertions as equally "delusional" or without credibility. She is misgendered with male pronouns, further invalidating her identity and creating distress. In the first excerpt, her behavior is characterized as "agitated, speech-loud" and in the second, her anger toward her father is presented as seemingly nonsensical. This suggests both that her assertions of abuse were not taken seriously and that the impact of abuse on her behaviors and emotions was pathologized. There was no further documentation of trauma or abuse in Irie's chart during her two-week stay. However, she was frequently characterized as "loud," "labile," "threatening," and "disruptive" and was maintained on close observation, suggesting the influence of anti-Black sanism in that she was perceived by practitioners as dangerous and threatening. Herman (2015) argues that

> [p]eople who have survived atrocities often tell their stories in a highly emotional, contradictory, and fragmented manner which undermines their credibility ... far too often secrecy prevails, and the story of the traumatic event surfaces not as a verbal narrative but as a symptom. (p. 1)

Irie provided fragmented information about her experiences of sexual violence, but because of the way she presented it and because anti-Black sanism and transmisogynoir made her less likely to be believed, the providers did not take it up and likely saw it as fantastical, despite the fact that childhood sexual abuse is highly correlated with the state of distress that gets called psychosis (Bentall et al., 2014; Read et al., 2014). This indicates the influence of rape culture in which sexual violence is invisibilized and Irie's behavior is seen as part of her psychotic illness rather than as a response to abuse.

What About the Perpetrators? Invisibilizing Acts of Sexual Violence

There is a lack of research data on perpetrators of sexual violence, including accurate numbers of how many are there, how and why they perpetrate violence, and their behaviors (Quadara, 2014). As Quadara (2014) argues, in media and in research, "the perpetrator slips quietly from view" (p. 41). What is known, however, is that the majority of perpetrators of sexual violence are cis men (Quadara, 2014). This trend is also observed in the current study, given that all of the charts with documentation regarding the perpetration of sexual violence are those of cis men.

Given the widespread nature of abuse and, therefore, the potentially large number of perpetrators in psychiatric hospitals, there is a "surprising lack of social or professional curiosity about the vast numbers of perpetrators who are helping to fill our psychiatric hospitals," notwithstanding the harm they cause (Johnstone, 2011, p. 107). This lack of interest is less surprising when we consider that rape culture leads to the scrutiny of survivors' actions and credibility rather than perpetrators' (don't get raped versus don't rape), while simultaneously normalizing and invisibilizing perpetrators' sexual violence. What happens when the gaze is turned on the charts of those who are possible perpetrators of sexual harassment and violence? Similar mechanisms of rape culture, sanism, and anti-Black sanism are evident in these charts, mirroring what is seen in the charts of survivors. Similar to how survivors' experiences of sexual violence are normalized, perpetrators' "behaviours, desires and interactions are normalised in and through social and systems' responses" (Quadara, 2014, p. 44).

For example, charts of perpetrators are filled with euphemistic language that invisibilizes and normalizes sexual violence. A white man who makes sexually violent statements to staff is characterized as "sexually frustrated." Another white man who provides detailed and specific plans to rape a female friend is likewise described as "sexually frustrated." A Black man is characterized as "sexually inappropriate" ("observed to be sexually inappropriate with female co-pts, touching/hugging"). In another example, rape is referred to as sex, when a Black man "believed his sister was his wife and he threatened to have sex with her." In light of anti-Black sanism, the use of "sexually frustrated" rather than "sexually inappropriate" is noteworthy, as it suggests that the behavior of white men is more understandable because of their "frustration" (if they could just have sex they wouldn't act this way, a belief informed by rape culture). The Black men are characterized as "inappropriate," a word that condemns their behavior but does not directly name it as sexual harassment. As I have discussed elsewhere (Pilling et al., 2018) and is further described in Zaheer's chapter (Chapter 5), the response to these displays of violence is racialized. In the examples given above, the first Black man who was noted as "sexually inappropriate" was noted in the same paragraph to have "required locked seclusion" in the past (thereby suggesting he was a threat) and the second Black man was placed on constant observation, whereas the general tone of the charts of the white men was more benevolent, and at times they were directly characterized as nonthreatening despite violent behavior. This shows the influence of anti-Black sanism on practitioners' documentation in the charts of Black men.

Sexual Violence as a Symptom of Psychosis

As discussed above, survivors' experiences of sexual violence are rarely meaningfully connected to distress and are often reconfigured as delusional. In contrast, perpetrators' acts of sexual violence are often directly linked to madness and seen as being caused by it, particularly in cases of psychotic illness. This phenomenon is gendered, in that it is most noticeable in the charts of cis men.[9] This conflates violence with madness, in that men's abusive behavior is never identified as an issue that is separate from mental distress, even in cases where it seems clear from reading the chart that the perpetrator has a long history of abusive behavior. This indicates the presence of rape culture and sanism as governing logics in the charts of perpetrators.

For example, it was documented numerous times that Ebo, a Black African, heterosexual, cis man had been physically and verbally abusive to his wife, including "punching," "pushing," and "beat[ing]" her. While there was no explicit documentation of sexual violence toward his wife, it is possible that this was less likely to be disclosed or fully documented. Indeed, a psychiatrist wrote about Ebo's "marital discord" in some detail, but then stated, "He spoke of more sensitive issues in their relationship, which writer will not detail here." This may or may not have been referring to issues of sexual violence. However, Ebo's chart contains many notes written by nurses and social workers that detail sexual harassment of women on the unit, both patients and staff, including unwanted touching, sexual comments, and looking in women's rooms. As noted above, euphemistic language is used to refer to this behavior. The abuse of his wife is described at various points as "aggression" and "hostility," while the sexual harassment on the unit is referred to as "sexually inappropriate" behavior.

Notably, Ebo's abusive behavior was interpreted by the psychiatrist as due to his alcohol use and his psychosis. For example, on day three of his admission, a psychiatrist wrote, "Dx: SCZP [Diagnosis: Schizophrenia] and more recent exquisite sensitivity to alcohol leading to psychosis exacerbation and aggressive b/h [behavior] towards his wife." In this example, a social issue (Ebo's abusive behavior toward his wife) was seen as a result of a supposedly biomedical issue (psychosis exacerbated by alcohol use). In other words, the psychiatrist saw Ebo's violence toward his wife as being due to his illness and a "sensitivity" to alcohol. Ten days later, Ebo went AWOL from the hospital, and a psychiatrist wrote the following:

> Although [Ebo] had been aggressive with his wife prior to his admission when experiencing psychotic Sx [symptoms], he has not exhibited that

> aggression on the unit nor has he ever made any threats or vocalized thoughts of hurting any of his family at all during this admission even during his more ill phase. He is likely not a risk to the family at this time having stabilized on medications but the team will continue to make efforts to contact the family and police will be notified accordingly.

This documentation stands in marked contrast to the documentation by other practitioners throughout the chart, in which Ebo is described as "sexually inappropriate" and "aggressive" with women (both patients and staff). Ebo's sexually violent behavior on the unit was thus invisibilized, and he is characterized as "not a risk to the family." It is clear that the psychiatrist believed that treating Ebo's psychosis with medication resolved his violent behavior toward his family, conflating madness with violence. The psychiatrist, therefore, seemed unconcerned for the safety of Ebo's wife and seemed to suggest that contacting the police was a formality. One of the results of viewing violence in this way is to assume, as the psychiatrist did, that biomedical treatment (medication in this case) will resolve violence. This approach allows violence to persist unchecked. In this example, Ebo returned to his wife's home while AWOL, despite the fact that he was aware that she was in the process of obtaining a restraining order and that he was not welcome there. In the documentation following Ebo's return to the hospital, the psychiatrist noted that his wife was away visiting family at the time, but did not indicate any concern about Ebo's decision to violate his wife's boundaries.

This example shows the problematic conflation of madness with cis men's abusive behavior, which is a sanist belief. While survivors' experiences of violence are rarely meaningfully connected to mental distress, perpetrators' acts of violence are rarely explicitly separated from it. And yet, the impact of violence and acts of violence are both pathologized as symptomatic of psychosis. For survivors, this means that they are disbelieved and the impact of their trauma is never meaningfully addressed, while for perpetrators, this means that their violent behavior is excused and unchecked. This phenomenon is governed by rape culture, which obfuscates and excuses acts of sexual violence while normalizing experiences of sexual violence and pathologizing the impact of sexual violence. This is also a gendered dynamic, in that all of the perpetrators were cis men, which reflects the larger societal reality that most perpetrators are cis men, notwithstanding the gender or sexuality of their victims (Armstrong et al., 2018).

Conclusion

This chapter raises questions about the responsibility of psychiatric institutions to address issues of sexual violence both in and outside of the institution. It is important to note that the experiences detailed were not "exceptional"; they are business as usual in this institution and society at large. Psychiatric hospitals do not currently address sexual violence and other forms of violence and oppression effectively, and this "institutional betrayal" can exacerbate distress for those who have experienced sexual violence (Smith & Freyd, 2013). Daley et al. (2011) raise questions about "the ethical responsibility of psych institutions and professionals to develop a political response to violence against women and children generally" (p. 9). Other scholars have described concrete suggestions regarding what this political response to violence could look like. For example, Humphreys and Thiara (2003) argue that the mental health system should shift away from the medical model toward approaches that meaningfully address violence. They suggest that mental health services should focus on naming violence, lifting blame, systematic proactive screening, safety planning, developing specialist support services for racialized and men survivors, and interventions focusing on experiences of violence (Humphreys & Thiara, 2003). Likewise, Godard et al. (2017) outline three phases for organizational change to implement a shift toward a violence- and trauma-informed lens in mental health and social services (pp. 434–435). The data for this study support these calls to action and broaden them to include the necessity of developing political responses to violence against trans people and recognizing the ways in which sexual violence and responses to it are imbricated with racism, anti-Blackness, sanism, and other forms of oppression. There is a need for approaches that challenge rather than uphold rape culture, sanism, and anti-Black sanism. This chapter suggests that psychiatric hospitals are not always the sites of respite and assistance they claim to be. When people go to psychiatric hospitals, they are experiencing very high levels of distress and, when they are there voluntarily, they are often desperately seeking help and support they have not been able to find elsewhere. In many cases, they are, instead, (re)victimized in myriad ways, including by new experiences of sexual violence, being disbelieved or pathologized based on their experiences of sexual abuse, being restrained and secluded (see Chapter 5), and being subject to discrimination as a matter of course such as misgendering, biphobia (see Chapter 4), sanism, and anti-Black sanism. This is because psychiatric institutions reinscribe colonial, white supremacist norms of gender, race, class, sexuality, and disability. And yet, the psychiatric institution is also a microcosm of society. In order to see profound change, we need a seismic shift in terms of the societal denial, normalization, and perpetuation of sexual violence, anti-Blackness, and other forms of violence and oppression.

Notes

1. I am politically aligned with critical scholarship and mad movement critiques of diagnoses and medicalizing labels such as "psychosis" to describe states of distress, while respecting self-identification with any term, label, or diagnosis. I employ the word psychosis here to indicate how the people whose charts are analyzed were interpellated by the psychiatric institution. Specifically, psychosis is defined in psychiatry as "conditions that affect the mind, in which people have trouble distinguishing between what is real and what is not." Symptoms may include "delusions ... hallucinations ... and disorganized speech." Some examples of diagnoses associated with psychosis include schizophrenia, bipolar disorder, and depression with psychotic features (https://www.camh.ca/en/health-info/mental-illness-and-addiction-index/psychosis).
2. My use of "sanism and anti-Black sanism" is not meant to reify a distinction between the two or to suggest that there is a type of sanism only experienced by non-Black people. As Ben-Moshe (2020) states "It's impossible to untangle antiblack racism from processes of pathologization, ableism, and sanism" (p. 25). I use the more cumbersome phrase "sanism and anti-Black sanism" because "sanism" has often been theorized in a white-centric way. Coined by Birnbaum (1960) and popularized by Perlin (1992), the term

"sanism" has been theorized in such a way that excludes the experiences of BIPOC. Anti-Black sanism, in particular, requires a term that explicates the "particularly perilous" compound of anti-Black racism and sanism that affects Black people (Meerai et al., 2016, p. 20).

3. Of note, a disproportionately high number of Black patients included in our sample were found in the unit designated for those with "serious mental illness," often characterized by "psychotic illness" (see Introduction–Methods).
4. All names given to patients are pseudonyms.
5. Vea's chart contained 43 instances of the word "obese" and eight instances of the word "overweight," indicating the presence of fatphobia in her chart.
6. For an in-depth, multifaceted consideration of this perspective, see the edited collection of Romme and Escher (2012).
7. The practice of copying and pasting verbatim sections of the chart is common. See Gibson (Chapter 2).
8. I use "sic" to indicate that Irie was misgendered by the practitioner.
9. This gendered dynamic refers specifically to documentation regarding sexual violence. A preliminary analysis suggests a similar logic in documentation regarding violence (that is not of a sexual nature) in the charts of cis women (i.e., acts of violence are conflated with madness, and "solved" with medication). This gendered dynamic in the charts regarding sexual violence likely exists because most perpetrators of sexual violence are cis men and because sexual violence between women is often not recognized as such (Daley & Ross, 2018). I cannot comment on what this dynamic may look like in the charts of trans patients as research team members have yet to conduct this analysis. For an analysis of the charts of trans patients as an independent data set, see Pilling (2022).

References

Abdillahi, I., Meerai, S., & Poole, J. (2017). When the suffering is compounded: Towards anti-Black sanism. In S. Wehbi & H. Parada (Eds.), *Reimagining anti-oppression social work practice* (pp. 109–122). Canadian Scholars Press.

Alcoff, L. M. (2018). *Rape and resistance*. Polity Press.

Armstrong, E. A., Gleckman-Krut, M., & Johnson, L. (2018). Silence, power, and inequality: An intersectional approach to sexual violence. *Annual Review of Sociology, 44*, 99–122. https://doi.org/10.1146/annurev-soc-073117-041410

Bailey, M., & Trudy. (2018). On misogynoir: Citation, erasure, and plagiarism. *Feminist Media Studies, 18*(4), 762–768. https://doi.org/10.1080/14680777.2018.1447395

Ben-Moshe, L. (2020). *Decarcerating disability: Deinstitutionalization and prison abolition*. University of Minnesota Press. https://doi.org/10.5749/j.ctv10vm2vw

Benn-John, J. (2016). (Re)colonizing Black Canadian women and routes to resistance. *Race, Gender & Class, 23*(1–2), 150–165. https://www.jstor.org/stable/26529195

Bentall, R. P., de Sousa, P., Varese, F., Wickham, S., Sitko, K., Haarmans, M., & Read, J. (2014). From adversity to psychosis: Pathways and mechanisms from specific adversities to specific symptoms. *Social Psychiatry and Psychiatric Epidemiology, 49*(7), 1011–1022. https://doi.org/10.1007/s00127-014-0914-0

Birnbaum, M. (1960). The right to treatment. *American Bar Association Journal, 46*(5), 499–505. https://www.jstor.org/stable/25721181

Daley, A., Costa, L., & Ross, L. E. (2011). Sexing up the psychiatric record: A pilot study of the documentation of women's sexuality in psychiatric inpatient charts. In G. Araoz (Ed.), *Madness in context: Historical, poetic and artistic narratives* (pp. 35–46). Brill.

Daley, A., & Ross, L. E. (2018). Uncovering the heteronormative order of the psychiatric institution: A queer reading of chart documentation and language use. In J. Kilty & E. Dej (Eds.), *Containing madness: Gender and "psy" in institutional contexts* (pp. 169–190). Palgrave Macmillan.

Donovan, R., & Williams, M. (2002). Living at the intersection: The effects of racism and sexism on Black rape survivors. *Women & Therapy, 25*(3–4), 95–105. https://doi.org/10.1300/J015v25n03_07

Fernando, S. (2010). *Mental health, race and culture*. Palgrave Macmillan.

Godard, L., Josewski, V., Cory, J., Abi-Jaoude, A., Malcoe, L. H., & Smye, V. (2017). Disrupting dominant discourses: Rethinking services and systems for women with experiences of abuse. In M. Morrow & L. Halinka (Eds.), *Critical inquiries for social justice in mental health* (pp. 430–459). University of Toronto Press.

Greco, D., & Dawgert, S. (2007). *Poverty and sexual violence: Building prevention and intervention responses*. Pennsylvania Coalition Against

Rape. https://pcar.org/resource/poverty-and-sexual-violence-building-prevention-and-intervention-responses. Accessed 12 December 2020.

Herman, J. (2015). *Trauma and recovery: The aftermath of violence—From domestic abuse to political terror*. Basic Books.

Humphreys, C., & Thiara, R. (2003). Mental health and domestic violence: "I call it symptoms of abuse". *British Journal of Social Work, 33*(2), 209–226. https://doi.org/10.1093/bjsw/33.2.209

Johnstone, L. (2011). Can traumatic events traumatize people? Trauma, madness and "psychosis." In M. Rapley, J. Moncrieff, & J. Dillon (Eds.), *De-medicalizing misery: Psychiatry, psychology and the human condition* (pp. 1–10). Palgrave Macmillan.

Kippert, A. (2016). *You can stop apologizing now: What it means when you can't stop saying "I'm sorry" after escaping abuse*. DomesticShelters.org. https://www.domesticshelters.org/articles/after-abuse/you-can-stop-apologizing-now. Accessed 15 December 2020.

Kumsa, M. K., Mfoafo-M'Carthy, M., Oba, F., & Gaasim, S. (2014). The contours of anti-Black racism: Engaging anti-oppression from embodied spaces. *Journal of Critical Anti-Oppressive Social Inquiry, 1*(1), 21–38.

Liegghio, M. (2013). A denial of being: Psychiatrization as epistemic violence. In B. LeFrancois, R. Menzies, & G. Reaume (Eds.), *Mad matters: A critical reader in Canadian Mad Studies* (pp. 122–129). Canadian Scholars Press.

McGuire, D. L. (2010). *At the dark end of the street: Black women, rape, and resistance—A new history of the civil rights movement from Rosa Parks to the rise of Black power*. Vintage Books.

Meerai, S., Abdillahi, I., & Poole, J. (2016). An introduction to anti-Black sanism. *Intersectionalities: A Global Journal of Social Work Analysis, Research, Polity, and Practice, 5*(3), 18–35.

Metzl, J. M. (2010). *The protest psychosis: How schizophrenia became a Black disease*. Beacon Press.

Miller, A. K. (2019). "Should have known better than to fraternize with a Black man": Structural racism intersects rape culture to intensify attributions of acquaintance rape victim culpability. *Sex Roles, 81*(7–8), 428–438. https://doi.org/10.1007/s11199-019-1003-3

Onwuachi-Willig, A. (2018). What about #UsToo: The invisibility of race in the #MeToo movement. *Yale Law Journal Forum, 128*, 105–120.

Perlin, M. L. (1992). On sanism. *SMU Law Review, 46*, 373–407.

Pilling, M. D. (2022). *Queer and trans madness: Struggles for social justice*. Palgrave Macmillan.

Pilling, M. D., Daley, A., Gibson, M. F., Ross, L. E., & Zaheer, J. (2018). Assessing "insight", determining agency and autonomy: Implicating social identities. In J. Kilty & E. Dej (Eds.), *Containing madness: Gender and "psy" in institutional contexts* (pp. 191–213). Palgrave Macmillan.

Powell, A., & Henry, N. (2014). Framing sexual violence prevention: What does it mean to challenge a rape culture? In N. Henry & A. Powell (Eds.), *Preventing sexual violence: Interdisciplinary approaches to overcoming a rape culture* (pp. 1–21). Palgrave Macmillan.

Preston, A. M. (2020, September 9). The anatomy of transmisogynoir. *Harper's Bazaar*. https://www.harpersbazaar.com/culture/features/a33614214/ashlee-marie-preston-transmisogynoir-essay/. Accessed 11 February 2021.

Quadara, A. (2014). The everydayness of rape: How understanding sexual assault perpetration can inform prevention efforts. In N. Henry & A. Powell (Eds.), *Preventing sexual violence: Interdisciplinary approaches to overcoming a rape culture* (pp. 41–62). Palgrave Macmillan.

Read, J., Bentall, R., & Fosse, R. (2014). Time to abandon the bio-bio-bio model of psychosis: Exploring the epigenetic and psychological mechanisms by which adverse life events lead to psychotic symptoms. In E. E. Speed, J. E. Moncrieff, & M. E. Rapley (Eds.), *De-medicalizing misery II: Society, politics and the mental health industry* (pp. 210–225). Palgrave Macmillan.

Romme, M., & Escher, S. (Eds.). (2012). *Psychosis as a personal crisis: An experience-based approach*. Routledge.

Slatton, B. C., & Richard, A. L. (2020). Black women's experiences of sexual assault and disclosure: Insights from the margins. *Sociology Compass, 14*(6), 1–12. https://doi.org/10.1111/soc4.12792

Smith, C. P., & Freyd, J. J. (2013). Dangerous safe havens: Institutional betrayal exacerbates sexual trauma. *Journal of Traumatic Stress, 26*(1), 119–124. https://doi.org/10.1002/jts.21778

Sweeney, B. N. (2020). Gender-based violence and rape culture. In N. Naples (Ed.), *Companion to women's and gender studies* (pp. 285–302). Wiley-Blackwell.

Tam, L. (2013). Whither Indigenizing the mad movement? Theorizing the social relations of race and madness through conviviality. In B. LeFrancois, R. Menzies, & G. Reaume (Eds.), *Mad matters: A critical reader in Canadian Mad Studies* (pp. 281–298). Canadian Scholars Press.

Unger, R. (2019). Moving toward a wider perspective on psychosis. *Journal of Humanistic Psychology, 59*(5), 768–774. https://doi.org/10.1177/0022167818762204

7

Concluding Thoughts

Margaret F. Gibson, Andrea Daley, and Merrick D. Pilling

This collection of writings challenges the common perception that the psychiatric chart is a neutral and disinterested text produced by individual actors. This project approaches the psychiatric chart as a mechanism that articulates institutional ideology and puts social relationships and discourses into action. More specifically, throughout the chapters of this book, the psychiatric chart is revealed as a text that stories

M. F. Gibson · A. Daley (✉)
School of Social Work, Renison University College, University of Waterloo, Waterloo, ON, Canada
e-mail: andrea.daley@uwaterloo.ca

M. F. Gibson
e-mail: margaret.gibson@uwaterloo.ca

M. D. Pilling
Women's and Gender Studies in the School of Social Work, University of Windsor, Windsor, ON, Canada
e-mail: Merrick.Pilling@uwindsor.ca

© The Author(s), under exclusive license to Springer Nature Switzerland AG 2021

A. Daley and M. D. Pilling (eds.), *Interrogating Psychiatric Narratives of Madness*,
https://doi.org/10.1007/978-3-030-83692-4_7

people and organizes them along the lines of gender, race, sexuality, and class.

In the psychiatric chart, we come face-to-face with the ways in which our larger society is storied and organized, most centrally along the lines of colonialism and white supremacy. Colonial and white supremacist ideology imbues hegemonic constructs of gender, race, sexuality, and class, heteronormativity and cisnormativity, monosexism, rape culture, sanism and anti-Black sanism, and racist and xenophobic tropes of the dangerous "Other." Accordingly, the contributions in this book illustrate that Black, Indigenous, and people of color (BIPOC), trans, poor, and gender transgressive people are especially vulnerable to the harm of being "marked" or erased through institutional sorting practices.[1]

All of the collection's contributors are especially concerned about the many examples of harm befalling BIPOC who experience emotional distress and encounter the psychiatric institution.[2] The very designation of "safety" as an institutional priority can produce further dangers for BIPOC people, who are often perceived as dangerous and met with institutional racism and other threatening forces rather than support (see Chapters 1, 3, 5, and 6). Sarah Ahmed (2015) notes that for racialized people, "institutional passing" requires self-defense through image management that makes the self unthreatening: "The experience of being a stranger in the institutions of whiteness is an experience of being on perpetual guard: of having to defend yourself against those who perceive you as somebody to be defended against" (para. 46).

While demands for police accountability in their too-often deadly encounters with BIPOC communities have sometimes led to calls for psychiatric services to supplant the policing role, the findings in this collection suggest that a simple substitution of one regulatory system for another is not nearly enough to create true safety for racialized people.

The contributors' analyses ask that we "keep the institution in view" (McCoy, 2006, p. 109). This requires more than constant questioning of policies or other institutional directives. When looking at psychiatric charting, this goal requires a refusal to be distracted by the supposed focus of the form—the individual patient and their momentary encounter. It requires that we "see" and challenge how ideology, structures, and histories shape psychiatric chart documentation and the ways in which ideology, structures, and histories are inscribed across time and space. Individual mental health professionals' (MHPs) documentation practices and patients cannot be the primary points of analysis. Nor can individual MHPs and institutions be the primary points of intervention. Once we see the porous replication and revision of textual discourse in and through institutional walls, we also see that the answer is not simply better institutional policy or new training for providers, although these may be helpful along the way. Rather, the ways in which gender, race, sexuality, and class structure Western institutions, including the psychiatric institution, and everyday encounters between patients and practitioners through chart documentation practices (i.e., institutional and professional forms) must be at the core of what we challenge—what we hope to change. Currently, these structures ask MHPs to notice and attend to different patients in different ways, along gendered, raced, sexualized, classed, and sanist lines, and with a generalized lens of suspicion. It is easy to be similarly caught in the framing of psychiatric charts as about one individual at a time rather than the coordination of people doing things in different spaces, at different moments—actions concerted by institutional(ized) ideology.

Writing—and reading—are acts of constrained imagination where a moment and/or an individual is rendered "as if" they are present and now. Charts are imagination by proxy. We imagine what happened to and with the individual named but through the perspective of someone who we are asked to imagine ourselves to be—a neutral (read white) person. As such, we are asked to imagine that if we were there at that time, we would also see and feel and understand in this way. The writer, in turn, is asked to imagine enough of the patient's world to explain whether and how it makes sense (i.e., to govern it) according to white, colonial, heterosexual, cis-gendered, classed, and neuronormative logic, and what risks and certainties it holds. But failures of imagination are inevitable, perhaps no more so when the acceptable lines of interpretation are laid out with standardized options and the writer's uncertainty is disallowed. When drop-down menus must be marked and categories assigned, possibilities are foreclosed; patients who report suspicions, traumas, and fears that are not readily imaginable and endorsed by the writer—or the dominant culture—are met with greater suspicion and judgment.

Based on the findings from this project, here are some suggested tactics for approaching psychiatric documentation in ways that avoid collapsing our attention onto what those designated as patients do and are, and instead focusing on the ways in which the institutional structures and relations that surround each document can be usefully interrogated.

1. We should challenge the institutional expectation that there is a right way to document or that institutional interests, provider interests, patient interests, and community interests can be collapsed into a single notion of professionalism. As long as the institutional role is closely integrated with colonial and white supremacist interests as they are reified through its relationship with medical, legal, and funding systems, there will be powers beyond the encounter shaping what happens—and what doesn't happen. Moreover, in this neoliberal context of auditing and liability, better documenting does not mean that people will be more fully, accurately, or respectfully understood and supported. Better charting in institutional terms means that there will be less variability and that fewer legal liabilities can be connected to what is documented/not documented. Colonial relationships of surveillance in the institution that are multi-layered and require that the charters be surveilled in how they surveil the patients are further entrenched in this context. We need to see charts as "happenings" that begin long before the encounter and end long after it and that hook the practices of individuals into the larger functions of social control as designated by the psychiatric institution and other institutions (e.g., police, courts, and child welfare).
2. The goals of charting need to be separate from the goals of the individuals contributing to the charts, or even the programs and organizations, in some cases. Mental health professionals are also constrained (although not necessarily harmed) by the processes and forms of charting—they too are subjected to the power of institutional ideology in which they work, are educated and trained, and are held to account. Charting requirements and taken-for-granted elements such as standardization and electronic forms that serve

as mechanisms for classification and order have consequences that extend far beyond any individual's intentions. The web of auditing and liability extends to regulatory colleges, courts, child welfare systems, and even public opinions (e.g., if something were to get into the media). People who chart and people who are charted have limited options as they encounter each other, and the need to chart/be charted for future possible readers and actions constrains all actors, although with much more proximate and likely outcomes for those charted (i.e., patients).

3. We need to ask whether, and where, colonial ways of knowing, being, and doing in the psychiatric institution can be challenged. Is there room to question the diagnostic classifications of the Diagnostic and Statistical Manual (American Psychiatric Association, 2013) as an "interested" text? Can we insist that social and structural violence be considered in a meaningful way in understanding and responding to distress? Is there room for shared and less hierarchical ways of being professionals? For example, it might be more accurate, open, and respectful to say, "I have no idea what this person's experience is like but I'm interested to learn what would be helpful to them." However, this is not an accepted charting practice or an approach to relationships that most MHPs are socialized to take. Mental health professions and professional education and training programs are themselves critical to the settler-colonial project. Psychiatry, nursing, and social work, for example, are similarly "circumscribed by the logics of conquest, extraction, apprehension, management, and pacification that advance the settler project and seek to secure settler futurity" (Fortier & Wong, 2019, p. 437; see also Canadian Association for Social Work Education, 2017; Czyzewski & Tester, 2014; Fernando, 2017; Lee & Ferrer, 2014; Mahone & Vaughn, 2007; Mcgibbon et al., 2013; Valderama-Wallace & Apesoa-Varano, 2020; Waiter & Nardi, 2019). Colonial interests embedded in health/mental health professions get in the way of uncertainty and encourage the imposition of a singular "expert" interpretation. The structure of the electronic psychiatric chart and professional training writes out practitioners, casting out the very social identities and locations that render

intelligible the distress narratives of patients that invoke social and structural injustice.

4. To this end, we might investigate the conditions that produce exceptions to institutional "practice as usual." For example, there is one documented exchange where a patient requests a set of items from a convenience store that the MHP (nurse) gets for them. This is an unusual kind of interaction to appear in the charts and shows a flexibility of role. The nurse noted that the patient was annoyed at how much the nurse spent, and they negotiated an arrangement where she returned some items and brought the change back. This kind of encounter might be normal in other settings, but it is worth investigating when and how these types of humane interactions that rupture (and thus reveal) the psychiatric institution as a site of colonial violence, albeit minimally, make it into the charts.

How we engage with each other and experience spaces and periods of our lives will always exceed language, let alone what can be written about us on a standardized and audit-ready chart. People whose names and words appear in psychiatric charts are always more than what they write or what is written about them. And yet what we see of these moments and how they are experienced can be understood by a turn to the charts and to a deeper understanding of how they are produced through the structures of normativity that undergird the psychiatric institution and what they, in turn, re-produce.

In navigating these documents, we can build on a metaphor from Dorothy Smith (1999), likening language to a map. As you read a map, you shift how you orient yourself and your next steps. Maps can get us to places and shape how we get there. They are graphic summaries, one way of representing what has happened over many messy and embodied moments, and they then act to coordinate the actions and thoughts of many people who may not even think about how these marks came to be. Maps are taken as the reality of the place and the route, and not a terribly interesting one at that. But maps, like charts, can be misleading, and the ways in which they mislead are often responsible for many repeated errors—harms—of those who come after.

However, we err as MHPs, educators in health/mental health professional education and training programs, researchers, and community activists if we start to see the map itself as the ultimate authority and the only reality of a space or a journey. We need to keep putting the institutions, with all their interpretations, conventions, limitations, and insights back into our understanding and our analysis. We need to put people and the occluded histories of their many actions back into how we read and write about charts. By paying attention to institutional structures and, relatedly, forms and conversations behind the writing and reading processes, talking with people (interviews), drawing connections through and across entries, and pulling in the histories of how people have been categorized, we move from texts as inanimate to texts as a part of how psychiatric institutions operate.

Notes

1. The language and requirements of psychiatry that make up the psychiatric chart are key to activating colonial and white supremacist ideology and the ways that difference, or transgression, is determined through institutionalized sorting processes.
2. BIPOC, trans, and poor people are especially likely to be brought into the institution through police and court involvement, which introduces a form of differentiation from the very beginning of an institutional engagement.

References

Ahmed, S. (2015). *Some striking feature: Whiteness and institutional passing*. Feministkilljoys. https://feministkilljoys.com/2015/06/14/some-striking-feature-whiteness-and-institutional-passing/. Accessed 20 December 2020.

American Psychiatric Association. (2013). *Diagnostic and statistical manual of mental disorders* (5th ed.). American Psychiatric Association.

Canadian Association for Social Work Education. (2017). *Statement of complicity and commitment to change*. https://caswe-acfts.ca/about-us/our-commitment-to-change/. Accessed 14 December 2020.

Czyzewski, K., & Tester, F. (2014). Social work, colonial history and engaging Indigenous self-determination. *Canadian Social Work Review, 31*(2), 211–226. http://www.jstor.org/stable/43486322

Fernando, S. (2017). *Institutional racism in psychiatry and clinical psychology: Race matters in mental health*. Palgrave Macmillan.

Fortier, C., & Wong, E. (2019). The settler colonialism of social work and the social work of settler colonialism. *Settler Colonial Studies, 9*(4), 437–456. https://doi.org/10.1080/2201473X.2018.1519962

Lee, E. O. J., & Ferrer, I. (2014). Examining social work as a Canadian settler colonial project: Colonial continuities of circles of reform, civilization, and in/visibility. *CAOS: The Journal of Critical Anti-Oppressive Social Inquiry, 1*(1), e1234p.

Mahone, S., & Vaughn, M. (Eds.). (2007). *Psychiatry and empire*. Palgrave Macmillan.

McCoy, L. (2006). Keeping the institution in view: Working with interview accounts of everyday experience. In D. Smith (Ed.), *Institutional ethnography as practice* (pp. 109–126). AltaMira.

Mcgibbon, E., Mulaudzi, F., Didham, P., Barton, S., & Sochan, A. (2013). Toward decolonizing nursing: The colonization of nursing and strategies for increasing the counter-narrative. *Nursing Inquiry, 21*(3), 179–191. https://doi.org/10.1111/nin.12042

Smith, D. E. (1999). Telling the truth after post-modernism. In *Writing the social: Critique, theory, and investigations*. University of Toronto Press.

Valderama-Wallace, C. P., & Apesoa-Varano, E. C. (2020). "The problem of the color line": Faculty approaches to teaching social justice in baccalaureate nursing programs. *Nursing Inquiry, 27*(3), e12349. https://doi.org/10.1111/nin.12349

Waiter, R., & Nardi, D. (2019). Nursing colonialism in America: Implications for nursing leadership. *Journal of Professional Nursing, 35*(1), 18–25. https://doi.org/10.1016/j.profnurs.2017.12.013

Glossary

Anti-Black sanism: Meerai et al. (2016) coined the term "anti-Black sanism," which they describe as "the historical and ongoing set of aggressions visited on Black/African people in the Global North" (p. 20). Abdillahi et al. (2017) state that Black/African people experience anti-Black sanism in mental health services related to "communication, diagnosis, hospitalization, treatment, intervention, and the involvement of the criminal justice system" (p. 122).

Antipsychotic medications: Medications used to treat symptoms of psychosis, including delusions and hallucinations.

Biphobia: Prejudice and discrimination against bisexual people. This can include erasure of bisexual identity or the belief that bisexuality is temporary or not real.

CBT: Cognitive-behavioral therapy. A type of therapy that is short term, goal oriented, and skills based. CBT is based on the linkage between one's thoughts, behaviors, and moods. Therapists guide patients to identify and re-examine thought processes to change problematic or undesired behaviors and moods.

Chemical restraints: Medical tranquilizers used to sedate patients, often those behaving aggressively or violently.

© The Editor(s) (if applicable) and The Author(s), under exclusive license to Springer Nature Switzerland AG 2021

A. Daley and M. D. Pilling (eds.), *Interrogating Psychiatric Narratives of Madness*,
https://doi.org/10.1007/978-3-030-83692-4

Cisgender: A person who identifies as the sex they were assigned at birth. "Cisgender" replaces terms such as "bio," "real," and "genetic," which suggest that trans people's gender identities are nonauthentic or unnatural.

Cisnormative: The belief in society that cisgender identify is the norm and standard of gender identity and, consequently, other gender identities are excluded, subjugated, or assumed not to exist.

Code white: An emergency code used in a hospital to signal and call for assistance. Needed in the event a patient is behaving violently.

Colonial ideology: The ideology rooted in colonialism and colonial conquest that rests on the social construction of white supremacy.

Coloniality: A unified system of control involving three levels of colonialism: Euro-American domination, the operation of subcolonial subjugations, and the hegemonic mind.

Depot (injection): A slow-release, injected form of antipsychotic medication. It is the same medication as tablet or liquid forms of antipsychotics, but its effects last much longer.

Duloxetine: An antidepressant medication used to treat depression and anxiety. It is in the class of SNRIs (see SNRIs).

Epistemic violence: The silencing of certain groups or individuals through limiting, denouncing, or preventing their communication, language, or knowledge-sharing. This must be understood in the context of colonialism, where one type of knowledge (Western) is privileged over another.

Euthymic: A "neutral" state of mood or behavior, between manic or depressive behavior, often used in reference to patients with bipolar disorder. This term implies that certain behavior is "normal" and other behavior, "abnormal."

Four-point restraints: Physical restraints used in healthcare settings that restrain a patient's arms and legs. They are most commonly used with patients who behave aggressively or violently toward hospital staff.

Fluoxetine: Often referred to by its commercial name *Prozac*, it is an antidepressant in the class of SSRIs used to treat mood disorders such as depression and anxiety.

Heteronormative: The belief in society that heterosexuality is the norm and standard of sexual orientation and, consequently other gender identities are excluded, subjugated, or assumed not to exist.

HS: *Hora somni*, meaning medication given at bedtime.

IM (medication) : Intramuscular. Medication that is injected directly into the muscles and reaches the bloodstream quickly.

MAOI: Monoamine oxidase inhibitors. An older form of antidepressant medications that are still considered effective by the medical community, but this class of medication is not often used because of the strict dietary requirements required by patients who are taking it.

Mechanical restraints: Physical restraints, such as handcuffs or straps, used with patients who behave aggressively or violently toward hospital staff.

Mirtazapine: An antidepressant medication.

Misogynoir: Anti-Black racism and misogyny directed specifically toward Black women and the Black woman experience. The term was first introduced in by Moya Bailey and Trudy (2018).

Monosexism: The assumption, belief, or value system that only single-gender sexual (i.e., attracted to no more than one gender) orientations (i.e., heterosexuality, homosexuality) are legitimate.

MSE: Mental status exam. An MSE is used to assess a patient's functioning and mental state. It is considered objective and analogous to a physical examination. Components of the MSE, in addition to insight and judgment, include appearance, attitude, behavior, level of consciousness, orientation, speech and language, mood, affect, thought process/form, thought content, suicidality and homicidality, and intellectual functioning.

NSG: Nurse or nursing.

NRT: Nicotine replacement therapy.

Olanzapine: A common antipsychotic medication.

PO: *Per os*, meaning medication taken orally (by mouth).

PRN: *Pro re nata*, meaning medication to be taken as needed, as prescribed by a doctor at their discretion. PRNs are often administered to patients to calm them down or to mildly sedate agitated or anxious patients.

PTSD: Posttraumatic stress disorder is a diagnosis given to people who respond with mental health symptoms to a traumatic experience of multiple traumatic experiences.

Sanism: The systematic subjugation and oppression of people who have received a mental health diagnosis or who are thought to be mentally ill.

Seclusion: In the context of psychiatry, isolating a patient from other patients by placing them in a room by themselves. Patients are secluded if they are at risk of harming themselves or others.

Seroquel: Antipsychotic medication often used to treat symptoms of schizophrenia such as delusions and hallucinations.

SNRI: Serotonin and norepinephrine reuptake inhibitors. A class of medication used to treat depression, anxiety, and chronic pain.

SSRI: Selective serotonin reuptake inhibitors. A class of medication used to treat depression and anxiety disorders.

STAT: *statim*, meaning immediately. Often used in a medical context to direct an action or instruction to be done immediately or shortly.

Suicidal ideation (SI): Thoughts or plans of suicide; can include either detailed plans or fleeting thoughts. Differs from suicidal attempt (SA), which is the action of suicide.

TCA: Tricyclic antidepressants. A class of antidepressant drugs.

Transgender: A person who does not identify with the sex they were assigned at birth and who may transition socially (name, pronouns, legal gender markers), physically (hair, clothes, binding, tucking, etc.), or medically (hormones, surgery).

Transmisogynoir: Prejudice, discrimination, or systemic oppression against Black transgender women.

Transmisogyny: Prejudice, discrimination, or systemic oppression against transgender women.

Transphobia: Negative attitudes and feelings about, and an aversion to, trans people. This can take the form of disparaging jokes, rejection, exclusion, denial of services, employment discrimination, name-calling, and violence.

Whiteness: A social construct that affords power and privilege to people who are/appear white.

White supremacist ideology: The belief in a society that whiteness and people who are/appear white are superior and the standard.

Zoloft: A drug in the class of SSRIs, commonly used to treat depression and anxiety disorders. Zoloft is the commercial name for sertraline.

References

Abdillahi, I., Meerai, S., & Poole, J. (2017). When the suffering is compounded: Towards anti-Black sanism. In S. Wehbi & H. Parada (Eds.), *Reimagining anti-oppression social work practice* (pp. 109–122). Canadian Scholars Press.

Bailey, M., & Trudy. (2018). On misogynoir: Citation, erasure, and plagiarism. *Feminist Media Studies, 18*(4), 762–768. https://doi.org/10.1080/14680777.2018.1447395

Centre for Addiction and Mental Health (CAMH). (2021a). *Antidepressant medications*. https://www.camh.ca/en/health-info/mental-illness-and-addiction-index/antidepressant-medications. Accessed 1 March 2021.

Centre for Addiction and Mental Health (CAMH). (2021b). *Antipsychotic medications*. https://www.camh.ca/en/health-info/mental-illness-and-addiction-index/antipsychotic-medication. Accessed 1 March 2021.

Centre for Addiction and Mental Health (CAMH). (2021c). *Posttraumatic stress disorder*. https://www.camh.ca/en/health-info/mental-illness-and-addiction-index/posttraumatic-stress-disorder. Accessed 1 March 2021.

Daniel M. S., & Crider C. J. (2003). Mental status examination. In M. Hersen & S. M. Turner (Eds.), *Diagnostic interviewing*. Springer. https://doi.org/10.1007/978-1-4757-4963-2_2

© The Editor(s) (if applicable) and The Author(s), under exclusive license to Springer Nature Switzerland AG 2021
A. Daley and M. D. Pilling (eds.), *Interrogating Psychiatric Narratives of Madness*, https://doi.org/10.1007/978-3-030-83692-4

DiAngelo, R. (2011). White fragility. *International Journal of Critical Pedagogy, 3*(3), 54–70.

Dotsen, K. (2011). Tracking epistemic violence, tracking practices of silencing. *Hypatia: A Journal of Feminist Philosophy, 26*(2), 236–257. https://doi.org/10.1111/j.1527-2001.2011.01177.x

Kortright, C. *Colonization and ideology*. https://theanarchistlibrary.org/library/chris-kortright-colonization-and-identity. Accessed 1 March 2021.

LeBlanc, S., & Kinsella, E. A. (2016). Toward epistemic justice: A critically reflexive examination of "sanism" and implications for knowledge generation. *Studies in Social Justice, 10*(1), 59–78. https://doi.org/10.26522/ssj.v10i1.1324

Martinot, S. (n.d.). *The coloniality of power: Notes toward de-colonization.* https://www.ocf.berkeley.edu/~marto/coloniality.htm. Accessed 3 February 2021.

Meerai, S., Abdillahi, I., & Poole, J. (2016). An introduction to anti-Black sanism. *Intersectionalities: A Global Journal of Social Work Analysis, Research, Polity, and Practice*, *5*(3), 18–35. https://journals.library.mun.ca/ojs/index.php/IJ/article/view/1682

Muir-Cochnrane, E., Grimmer, K., Gerace, A., Bastiampillai, T., & Oster, C. (2019). Prevalence of the use of chemical restraint in the management of challenging behaviours associated with adult mental health conditions: A meta-synthesis. *Journal of Psychiatric and Mental Health Nursing, 27*(4), 425–445. https://doi.org/10.1111/jpm.12585

Polania Gutierrez, J. J., & Munakomi, S. (2020). *Intramuscular injection.* National Center for Biotechnology Information. https://www.ncbi.nlm.nih.gov/books/NBK556121/. Accessed 1 March 2021.

QMUNITY. (2018). *Queer terminology from A to Q*. https://qmunity.ca/wp-content/uploads/2019/06/Queer-Glossary_2019_02.pdf. Accessed 1 March 2021.

Ogbru, A., & Marks, J. W. (n.d.). *Tricyclic antidepressants (TCAs).* RxList. https://www.rxlist.com/tricyclic_antidepressants_tcas/drug-class.htm. Accessed 1 March 2021.

Rector, N. A. (2010). *Cognitive-behavioural therapy: An information guide.* Centre for Addiction and Mental Health.

University Health Network. https://www.uhn.ca/corporate/AboutUHN/Governance_Leadership/Policies/Documents/5.80.001_Emergency_Preparedness_Code_White_Violent_Persons.pdf. Accessed 1 March 2021.

Index

© The Editor(s) (if applicable) and The Author(s), under exclusive license to Springer Nature Switzerland AG 2021
A. Daley and M. D. Pilling (eds.), *Interrogating Psychiatric Narratives of Madness*,
https://doi.org/10.1007/978-3-030-83692-4

C

urer's authorised representative in the EU is Springer
Service Centre GmbH, Europaplatz 3, 69115 Heidelberg,
e any concerns regarding our products, please
@springernature.com

oup (UK) Ltd, Croydon, CR0 4YY
026
003